U0473093

THE SECOND 6 YEARS OF IRCK

The International Research Centre on Karst (IRCK) under the Auspices of UNESCO
Institute of Karst Geology, Chinese Academy of Geological Sciences

SCIENCE PRESS
BEIJING

国际岩溶研究中心第二个六年历程

联合国教科文组织国际岩溶研究中心
中国地质科学院岩溶地质研究所 编著

科学出版社
北京

内 容 简 介

联合国教科文组织国际岩溶研究中心第二个六年历程从组织建设与管理、科学研究、国际交流与培训、科学普及与咨询等四个方面对中心第二个运行期内的工作进行了全面的总结，系统展示了中心在履行中国政府与联合国教科文组织签署的协定中所做的系列工作，为推动岩溶科学发展，为支撑中国政府国家战略、联合国教科文组织优先战略及部门优先计划等付出的努力。本书既是对已有辛勤耕耘的记录，亦是对未来发展的驱策。

本书面向全球范围内关注岩溶事业发展的科研工作者、政府决策者以及社会参与者，旨在通过国际合作为岩溶区可持续发展政策的制定提供科学支撑，为大众认识岩溶、提升岩溶保护意识做出贡献。

审图号：GS京（2023）1522号

图书在版编目(CIP)数据

国际岩溶研究中心第二个六年历程/联合国教科文组织国际岩溶研究中心，中国地质科学院岩溶地质研究所编著. —北京：科学出版社，2023.12

ISBN 978-7-03-074028-1

Ⅰ. ①国⋯ Ⅱ. ①联⋯ ②中⋯ Ⅲ. ①岩溶－研究机构－概况－世界 Ⅳ. ①P642.25-241

中国版本图书馆CIP数据核字(2022)第228839号

责任编辑：周 丹 沈 旭 / 责任校对：樊雅琼
责任印制：张 伟 / 封面设计：许 瑞

科学出版社 出版
北京东黄城根北街16号
邮政编码：100717
http://www.sciencep.com

北京中科印刷有限公司印刷
科学出版社发行　各地新华书店经销

*

2023年12月第　一　版　开本：889×1194　1/16
2023年12月第一次印刷　印张：24
字数：800 000

定价：599.00元

（如有印装质量问题，我社负责调换）

THE SECOND 6 YEARS OF IRCK Editorial Board

Editor: Peng Xuanming

Subeditors: Cao Jianhua   Luo Qukan

Technical Advisers: Yuan Daoxian   Jiang Zhongcheng

Translations and Editions: Luo Qukan   Bai Bing   Zhong Liang

Members of Editorial Board: (Sequencing in Pinyin of name order)

Bai Bing	Chen Hongfeng	Chen Weihai	Deng Zhenping	Dong Jige
Gan Fuping	Huang Baojian	Lei Mingtang	Li Jie	Li Qiang
Li Wenli	Liang Yongping	Liu Ning	Luo Weiqun	Luo Xuebing
Pan Xiaodong	Qin Xiaoqun	Su Chuntian	Su Luxuan	Wu Qing
Xia Riyuan	Xu Qi	Yang Chuchang	Zhang Cheng	Zhang Jing
Zhang Ling	Zhang Yuanhai	Zhao Xiaoming	Zhong Liang	Zhong Kaiwei
Zhou Lixin	Zou Shengzhang			

《国际岩溶研究中心第二个六年历程》编委会

主　　编：彭轩明

副 主 编：曹建华　罗劬侃

技术顾问：袁道先　蒋忠诚

编　　译：罗劬侃　白　冰　钟　亮

编　　委：（按姓名汉语拼音排序）

白　冰	陈宏峰	陈伟海	邓振平	董继革
甘伏平	黄保健	雷明堂	李　杰	李　强
李文莉	梁永平	刘　宁	罗为群	罗雪兵
潘晓东	覃小群	苏春田	苏橹萱	吴　庆
夏日元	许　琦	杨初长	章　程	张　晶
张　玲	张远海	赵小明	钟　亮	钟开威
周立新	邹胜章			

Foreword

On 12 May 2016, the government of the People's Republic of China and UNESCO successfully signed *the Agreement between the Government of the People's Republic of China and the United Nations Educational, Scientific and Cultural Organization Concerning the International Research Centre on Karst in Guilin, China, under the Auspices of UNESCO (Category 2)*, the International Research Centre on Karst (IRCK) officially entered the second six-year operation.

During the second six-year operation, IRCK has been provided with human resources, material, financial and governmental support by UNESCO and the Chinese government, especially the Sector of Natural Sciences of UNESCO, Ministry of Natural Resources, Ministry of Science and Technology, Ministry of Foreign Affairs of China, National Natural Science Foundation of China, China Geological Survey, Chinese Academy of Geological Sciences, Department of Science and Technology of Guangxi and Guilin Science and Technology Bureau; moreover, IRCK has also received technical support from the Governing Board, Academic Committee, related organizations and scientists from relevant countries.

Achievements on scientific research. On 14 November 2016, the announcement ceremony of the International Big Scientific Plan on "Resources and Environmental Effects of Global Karst Dynamic Systems" (Global Karst) was hosted by China Geological Survey in Guilin with high attention paid by the international karst academic community. Irina Bokova, director general of UNESCO, and Jiang Daming, minister of the Ministry of Land and Resources of China, sent congratulatory letters. In September 2019, the first-phase progress of Global Karst was officially released in the form of an international seminar. During the second-phase operation of IRCK, gratifying progress about Global Karst has been made:

● The research achievements on karst carbon sink, with great attention paid by the Government of China, were listed by the carbon peaking and carbon neutrality goals. The first natural resources industrial standard on karst carbon sink was released. An interesting discovery is that the karst cave is a favorable environment for methane degradation.

● Developed the ecological restoration technology for regulating water-soil-plant resources in karst rocky desertification areas, and established a karst-featured ecological industrial chain. In 2018, a briefing titled as *A Significant Progress in the Control of Karst Rocky Desertification in China* was reported to the Ministry of Science and Technology.

● Revealed the formation mechanism and classification of karst collapse, and raised the related countermeasures to geohazards risks evaluation and prevention, based on perfecting the monitoring and early warning system of karst collapse.

● Improved the evaluation models on karst groundwater hydrological process and water resources, and enhanced the karst groundwater detection technology. Generated the modes for optimizing the regulation and utilization of karst groundwater in the Xiaojiang River Basin of Luxi, Yunnan Province, with the utilization efficiency of water resources increased by 25%.

● Supported Xiangxi in China and Satun in Thailand to apply for the UNESCO Global Geopark successfully. Supported Wulong World Natural Heritage Site in Chongqing, China, to develop sustainably.

What is more delightful is that during the second operation period, a number of bilateral and multilateral international cooperative projects were successfully approved and implemented by joint efforts from different parties. IRCK participated in the work of the *World Karst Aquifer Map*; compiled the monograph of *Karst Geology and Development within Tethys Domain*; cooperated with the scientists from ASEAN countries to analyze the types of karst water storage, water environment problems, and technical solutions in Southeast Asia, with a relevant paper published jointly. In cooperation with the Chinese National Committee for Man and the Biosphere Programme of UNESCO, the popular science reading *Karst Geology and Ecosystem* was completed. In 2016, IRCK successfully recommended Professor Chris Groves of Western Kentucky University from USA to win the International Scientific and Technological Cooperation Award of the People's Republic of China.

Achievements on platform operation. Since 2020, COVID-19 has impacted academic exchanges globally. The secretariat of IRCK has adhered to the highly efficient operation, overcame difficulties, and explored new ways to forward.

During the second operation period, IRCK held 3 Governing Board meetings (2 on-site, 1 online), 3 academic committee meetings (2 on-site, 1 online), and 8 International Training Courses (6 on-site, 2 online). Considering the time lag between different countries, the online training courses were organized in the morning or afternoon of Beijing time respectively, attracting 228 students from 44 countries of 4 continents, and inviting 26 karst experts and scholars from 15 countries as lecturers. Among them, there are 220 from developing countries, and 30 from African countries. The training courses integrated with the UNESCO family better than ever. In 2017, the representative of the Permanent Delegation of the People's Republic of China to UNESCO was invited as the lecturer. In 2021, Dr. Ozlem Lopes from the Division of Ecological and Earth Sciences of UNESCO was invited as the lecturer. In 2020, the UNESCO Beijing Office helped IRCK to enroll the trainees. Moreover, in 2021–2022, IRCK participated in the activities or organized serial scientific and public activities for the International Year of Caves and Karst sponsored by the International Union of Speleology actively.

The training continued its improvement. Mr. Chaiporn Siripornpibul from Thailand participated in the training as a trainee in 2015 and was invited to serve as the lecturer in 2018–2020. He now acts as the leading scientist on karst of a Thai karst resources and environment innovation team. Mr. Eko Haryono, Indonesia, participated in the training as a trainee from 2009 to 2010 and was invited to serve as the lecturer in 2011, 2016, 2018, and 2019. He now acts as the leading scientist of the Indonesian innovation team on karst hydrology and geomorphology. In 2020, he started his role as the chair of the Karst Commission under the International Geographical Union, a crucial position to spread karst science. Ms. Liza Manzano from the Philippines participated in the training courses as a trainee in 2019 and was invited as a lecturer in 2020. She now acts as the leading scientist of the innovation team on karst landscape protection and disaster prevention. In 2021, by cooperating with the three teams, IRCK successfully applied the international cooperation project of "Guangxi–ASEAN Karst Landscape Resources Sustainable Utilization R&D Demonstration Platform" granted by the Department of Science and Technology, Guangxi.

Mr. Harrison Pienaar from South Africa was invited as the lecturer of the training course in 2013 and 2019, who acts as the leading scientist of the karst water resources research team of South Africa. Mr. Innocent Muchingami from Zimbabwe participated in the training courses as a trainee in 2013, 2014, and 2018, who now acts as the leading scientist of the karst hydrogeological research team of the National University of Science and Technology (NUST) in Zimbabwe. Moreover, NUST signed a cooperative agreement with IRCK. From 2016 to 2018, The China–Africa Water Forum was successfully held in China, Zimbabwe, and Egypt respectively.

IRCK maintained a good cooperation with Karst Research Institute ZRC SAZU of Slovenia, with its staff Mr. Tadej Slabe, Mr. Martin Knez, Ms. Nataša Ravbar, and Mr. Mitja Prelovšek as the lecturers of the training courses in last six years. In addition, IRCK cooperated with Slovenia side to apply for the Inter-governmental S&T Cooperation Project on "Infiltration Processes through the Unsaturated Zone of Karst Aquifers in Temperate and Subtropic Climates" successfully with 3 papers published jointly.

In terms of platform operation during the second six years, it is particularly worth mentioning that IRCK successfully applied a Technical Committee on Karst of the International Organization for Standardization (ISO/TC 319) in 2018. In 2020, IRCK successfully joined the International Union of Geological Sciences as its affiliated organization; in 2021, it successfully joined the Group on Earth Observation as its participating organization. The functions of IRCK were fully played.

Finally, I would like to extend my congratulations on the achievements made by IRCK during the second-phase operation and the publication of this book, in a speech delivered by Xi Jinping, president of the People's Republic of China at UNESCO Headquarters in 2014:

We should develop science and technology more vigorously. Scientific advancement and innovation can help people understand themselves and the world and be in a stronger position to change their society for the better. The continued process of exploiting nature will enable our people to master still more knowledge and skills. In this sense, science and technology are a powerful tool to make the world a better place for mankind.

Cao Jianhua
The executive deputy director of UNESCO International Research Centre on Karst

序

2016年5月12日，中华人民共和国政府与联合国教科文组织（以下简称"教科文"）在河北廊坊顺利签署了《中华人民共和国政府与联合国教育、科学及文化组织关于在中国桂林建立由教科文组织支持的国际岩溶研究中心（第2类）的协定》（以下简称《协定》），国际岩溶研究中心（以下简称"中心"）正式步入第二个六年运行期。

中心在第二个六年运行期内，得到教科文和中国政府，尤其是教科文自然科学部和中国自然资源部、科技部、外交部、国家自然科学基金委员会，以及中国地质调查局、中国地质科学院和广西壮族自治区科学技术厅、桂林市科学技术局等部门提供的人力、物力、财力和组织保障；得到中心理事会、学术委员会及相关国家的科学家和机构提供的技术支持。

科学研究取得成效。 2016年11月14日，"全球岩溶动力系统资源环境效应"（以下简称"全球岩溶"）国际大科学计划在中国地质调查局主持下及国际岩溶学界的高度关注下在桂林正式启动，教科文组织总干事伊琳娜·博科娃、国土资源部部长姜大明发来贺信；2019年9月，"全球岩溶"国际大科学计划第一阶段成果通过国际研讨会正式对外发布。在中心第二期运行期间，该国际大科学计划所涉及的研究领域均取得可喜进展：

● 岩溶碳汇研究成果得到中国政府的高度重视，列入国家"双碳"目标中，第一部有关岩溶碳汇的行业标准颁布；其中一个很有趣的发现是岩溶洞穴环境是甲烷降解的优越环境。

● 研发岩溶石漠化区水–土–植资源调控生态修复技术，构建岩溶特色生态产业链；2018年，形成科技部简报《我国喀斯特石漠化治理取得重大进展》。

● 在完善岩溶塌陷监测、预警系统的基础上，揭示岩溶塌陷的形成机制、类型划分，提出地质灾害风险评价和预防对策。

● 发展了岩溶地下水文过程、水资源评价模型，提高了岩溶地下水探测技术，形成了云南泸西小江流域岩溶地下水优化调控、高效利用模式，水资源利用效率提高 25%。

● 助力中国湘西、泰国沙敦成功申报世界地质公园；助力重庆武隆世界自然遗产地可持续发展。

更可喜的是中心在第二个运行期，通过多方努力，一批双多边国际合作项目成功获批并执行；中心参与编制了《全球岩溶含水层图》；编写了《特提斯域岩溶地质与岩溶发育》专著；联合东盟国家科学家，对东南亚岩溶水赋存类型、面临的水环境问题及解决的技术途径进行了分析，并共同发表相关成果；与中华人民共和国人与生物圈国家委员会合作，编著了科普读物《岩溶与生态》。2016 年，中心成功推介美国西肯塔基大学克里斯·葛立夫教授荣获中华人民共和国国际科学技术合作奖。

平台管理取得成效。2020 年以来，新冠疫情给国内外学术交流带来了一定的影响，中心秘书处秉承了高效运作的机制，克服困难，探索前行。

中心在二期运行期内，举办理事会 3 次（线下 2 次、线上 1 次）、学术委员会 3 次（线下 2 次、线上 1 次）。举办 8 届国际培训班（线下 6 届，线上 2 届），考虑到不同国家间的时差，线上培训采用北京时间上午班和下午班，吸引了四大洲 44 个国家的 228 名学员参加，邀请了 15 个国家 26 位岩溶专家、学者作为教员。其中发展中国家学员 220 名，来自非洲国家的学员 30 名；培训班进一步融入了教科文大家庭，2017 年培训班邀请了中国驻教科文常驻团的专家作为教员，2021 年邀请教科文生态地学部奥兹莱姆·洛佩斯博士作为教员，2020年培训班得到了教科文组织驻华代表处的宣传推广。此外，2021~2022年，中心积极参与国际洞穴联合会发起的国际洞穴与岩溶年活动，举办了系列科学和公益活动。

　　培训效果持续向好，泰国柴鹏·斯里蓬皮布尔于2015年作为学员参加培训，后受邀于2018~2020年担任培训班教员，以其为领军人才形成了泰国岩溶资源环境创新团队；印度尼西亚艾可·哈约诺于2009~2010年作为学员参加培训，后受邀于2011年、2016年、2018年、2019年担任培训班教员，以其为领军人才形成了印度尼西亚岩溶水文地貌创新团队，并于2020年担任国际地理联合会岩溶专业委员会主席，实现了东盟岩溶地质科技人才国际话语权的历史性突破；菲律宾丽萨·曼赞诺于2019年、2020年先后以学员和教员身份参加培训班，以其为领军人才形成了菲律宾岩溶景观保护与灾害防治创新团队；2021年，该三支岩溶地质科技创新团队，与中心合作，成功获批广西科学技术厅支持的"广西–东盟岩溶景观资源可持续利用研发示范平台"国际合作项目。

　　南非哈里森·皮纳尔于2013年、2019年担任培训班教员，以其为领军人才形成了南非岩溶水资源研究团队；津巴布韦因纳森·穆钦加米于2013年、2014年、2018年以学员身份参加培训，以其为领军人才形成了津巴布韦国立科技大学（NUST）的岩溶水文地质研究团队，并且NUST与中心签署了合作协议，2016~2018年分别在中国、津巴布韦、埃及成功举办中–非水资源论坛。

　　中心与斯洛文尼亚岩溶研究所一直保持良好的合作关系，其中特德·斯莱布、马丁·内兹、娜塔莎·阿尔巴、米提亚·普利罗斯克等，担任了六届培训班教员，并成功联合申报了"温带与亚热带岩溶含水层包气带渗流过程及气候环境意义"政府间国际科技合作项目,共同发表学术论文3篇。

在第二期平台管理成效方面,特别值得一提的是,中心2018年成功申请并设立了国际标准化组织岩溶技术委员会（ISO/TC 319）；2020年成功加入国际地质科学联合会,成为其附属机构；2021年成功加入地球观测组织,成为其参与机构,中心的职能得以充分发挥。

最后,想用2014年中华人民共和国国家主席习近平在联合国教科文组织总部演讲中的一句话,来表达我对中心在第二期运行期间取得的成效和本书的出版表示祝贺：

我们要大力发展科技事业,通过科技进步和创新,认识自我,认识世界,改造社会,使人们在持续的天工开物中更好掌握科技知识和技能,让科技为人类造福。

联合国教科文组织国际岩溶研究中心常务副主任 曹建华

目录
CONTENTS

Chapter 1　General Introduction

003　第一章　概况

Chapter 2　Organization and Management

015　第二章　组织建设与管理

Chapter 3　Scientific Research

073　第三章　科学研究

Chapter 4 International Exchange and Training

191　第四章　国际交流与培训

Chapter 5 Scientific Popularization and Consultation

297　第五章　科学普及与咨询

Chapter 6 Summary

327　第六章　总结

Fengcong karst is defined as a group of rocky hills or peaks rising from shared limestone foot-slopes. Closed depressions lie between the peaks, so the landscape is sometimes described as peak cluster depression. Fenglin karst is defined as limestone towers or peaks that are isolated from each other by flat limestone surfaces, which are generally covered by a thin layer of loose sediment. The peaks are usually completely surrounded by a karst plain, so the landscape may be called a peak forest plain. Guilin features world-class fenglin-fengcong karst landscape.

峰丛地形指具有连座的一些石峰，其间常有封闭洼地，因此其组合形态也可称为峰丛洼地。狭义的峰林地形是指被一片平地所分割的一些石峰，其组合形态可以分为峰林平原、峰林谷地、峰林盆地等。桂林是岩溶峰林、峰丛最为典型的代表。

（本页照片由朱学稳提供）/ This photo is from Zhu Xuewen

Chapter 1

General Introduction

第一章 概况

On 12 May 2016, the International Research Centre on Karst under the auspices of UNESCO stepped into the second six-year operation formally. By focusing on *The Agreement between the Government of the People's Republic of China and the United Nations Educational, Scientific and Cultural Organization Concerning the International Research Centre on Karst in Guilin, China, under the Auspices of UNESCO (Category 2)* (the Agreement), *UNESCO's Approved Programme and Budget (38C/5)*, and *The 2030 Agenda for Sustainable Development*, IRCK has made more systematic and outstanding achievements in the organization and management, scientific research, and social service, international exchange and training, as well as science popularization and consultation.

2016年5月12日,联合国教科文组织国际岩溶研究中心(以下简称"中心")正式步入第二个六年运行期。中心围绕《中华人民共和国政府与联合国教育、科学及文化组织关于在中国桂林建立由教科文组织支持的国际岩溶研究中心(第2类)的协定》《教科文组织批准的计划与预算(38C/5)》《2030年可持续发展议程》,在组织管理、科学研究与社会服务、国际交流与培训、科学普及与咨询方面均取得了更加系统、更具突破性的进展。

1.1 The Objectives and Functions of the Renewal Agreement
二期协定目标与职能

On 12 May 2016, Mr. Flavia Schlegel, the ADG of UNESCO (row 1, the second from left), and Mr. Cao Weixing, the vice minister of the Ministry of Land and Resources (row 1, the second from right) signed the renewal agreement on IRCK. Compared with the first phase, IRCK has set more specific objectives and functions.

2016年5月12日，联合国教科文组织助理总干事弗莱维娅·斯莱格尔女士（前排左2）与国土资源部代表曹卫星副部长（前排右2）签署了中心二期协定，与一期协定相比，中心设定了更明确的目标与职能。

1. Objectives
中心目标

(a) Advance knowledge on karst dynamic systems through scientific research, publications, international cooperation.

通过科学研究、出版活动和国际合作，促进岩溶动力学发展。

(b) Advance international cooperation and contacts and provide a platform for the exchange of scientific information about karst dynamic systems and the sustainable utilization of karst resources and environmental protection, between institutions worldwide within the framework of the International Geosciences and Geoparks Programme (IGGP) of UNESCO, the karst commissions of the International Union of Geological Sciences (IUGS), the International Association of Hydrogeologists (IAH), and the International Geographical Union (IGU).

促进国际合作与交往，为世界各地属于教科文组织国际地球科学与地质公园计划（IGGP）、国际地质科学联合会（IUGS）、国际水文地质学家协会（IAH）以及国际地理联合会（IGU）岩溶相关委员会在内的机构搭建一个有关岩溶动力系统、岩溶资源可持续利用和生态环境保护的科学信息交流平台。

(c) Provide advisory activities, technical information and training, and raise social awareness on karst dynamic systems applications for integrated control of rocky desertification rehabilitation, ecological restoration and biodiversity, and global climate change, including society at large, NGOs, and governmental institutions at central and regional levels.

提供咨询服务、技术信息和培训，提高社会（包括社会大众、非政府组织以及中央和地方的政府机构）对岩溶石漠化恢复综合治理、生态恢复、生物多样性及全球气候变化的认识。

(d) Develop a network of demonstration sites for the implementation of karst dynamic systems theory to improve epikarst and subterranean water resources utilization, control rocky desertification rehabilitation, protect biodiversity to improve its capacity to adapt to climate change, study the karst process and carbon cycle, and examine the paleo-climate record with stalagmites, so as to provide services for karst areas' sustainable development.

为岩溶动力系统理论的实践，建立示范基地网，主要包括提高岩溶表层水、地下水的资源利用率，石漠化治理，保护生物多样性提高其适应气候变化能力，岩溶作用与碳循环，洞穴石笋对古气候变化记录等，为岩溶地区可持续发展提供服务。

(e) Promote monitoring, modeling, and mapping systems on karst dynamic systems, gradually put into place a monitoring network and conduct comparative studies among different countries and regions.

促进建立岩溶动力系统监测系统、建模系统和相关制图系统，逐步构建监测网络，开展不同国家和地区对比。

2. Functions

中心职能

(a) Conduct technical, scientific, and theoretical research on modern karst science; coordinate and organize international/interregional cooperative projects and provide guidance on experimental equipment and provide field experimental sites when necessary.

开展现代岩溶学方面的技术、科学和理论研究；协调并组织国际及地区间的合作项目，在必要时提供实验设备使用指导和现场实验基地。

(b) Revise and update the website of the Centre, and build a platform to facilitate the exchange of scientific and policy information at the international level; undertake science popularization and dissemination, actively organize and participate in international workshops, symposiums and conferences related to karst science.

修改并更新中心网页，搭建交流平台以便开展国际科技信息和政策信息的交流；开展科普活动，积极举办和参与与岩溶相关的国际专题研讨会、国际专题学术会议和综合性国际学术会议。

(c) Cooperate with the International Geoscience and Geoparks Programme (IGGP), the International Hydrological Programme (IHP), the Man and the Biosphere (MAB) Programme, government agencies, NGOs, and decision-makers, in order to put the results of scientific research into practice.

与国际地球科学与地质公园计划（IGGP）、国际水文计划（IHP）、人与生物圈计划（MAB）、政府机构、非政府组织和决策人合作，以将科研成果转化为实际生产力。

(d) Organize international training courses on karst science for postgraduate students, researchers, technicians and relevant managers, provide up-to-date progress, technologies and policies related to the karst dynamic system, arrange field visits, and promote in-depth exchange among participants.

举办研究生、科技人员和相关管理人员的国际岩溶培训班，提供最新的岩溶动力系统研究进展、技术和相关政策，安排野外考察，促进深入交流。

(e) Gradually develop guidelines and criteria on the investigation and research on karst dynamic systems, conduct comparative study, advance ecological education, rocky desertification rehabilitation, and biodiversity protection, address global climate change and sustainable development to raise public awareness.

逐步制定岩溶动力系统调查和研究的指南与准则，开展对比研究，推进生态教育、石漠化治理恢复、生物多样性保护、应对全球气候变化和可持续发展，以提高社会公众意识。

Through a 6-year operation, IRCK has reached the objectives and performed the functions set by the Agreement through scientific research, publications, social service, international training, international exchange, and science popularization. Moreover, by kicking off the International Big Scientific Plan on "Resources and Environmental Effects of Global Karst Dynamic Systems", IRCK fostered the profound development of karst science further, promoted the mutual exchange between IRCK and other international organizations, and realized the overall objective that karst science should support the sustainable development of karst areas.

通过六年运转，中心通过科学研究、出版活动、社会服务、国际培训、国际交流、科学普及，较好地完成了各项目标与职能。此外，通过启动"全球岩溶动力系统资源环境效应"国际大科学计划，推进了岩溶科学研究的深远发展，推进了中心与国际组织间的互动交流，逐步实现了岩溶科学服务岩溶区可持续发展的既定目标。

1.2 UNESCO's Approved Programme and Budget (38C/5)
教科文组织批准的计划与预算（38C/5）

In 2015, the UNESCO Executive Board decided to renew the status of IRCK as a Category 2 Centre in 197ex/16 (I.A) Resolution. According to the approved Programme and Budget (38C/5) approved by UNESCO in 2015, the priority areas for the natural science sector are as follows:

2015年，教科文组织执行局在第197ex/16 (I.A)号决定中决定续延国际岩溶研究中心作为第2类中心的地位。根据2015年度《教科文组织批准的计划与预算（38C/5）》，其中关于自然科学部门的优先领域如下：

<table>
<tr><th colspan="5">Major Programme II
重大计划 II</th></tr>
<tr><td rowspan="2">37C/4 Strategic Objectives
战略性目标</td><td colspan="2">SO 4:
Strengthening science, technology and innovation systems and policies nationally, regionally and globally
战略性目标 4：
在国家、地区和全球层面加强科技与创新体系和政策</td><td colspan="2">SO 5:
Promoting international scientific cooperation on critical challenges to sustainable development
战略性目标 5：
促进国际科学合作，应对可持续发展的主要挑战</td></tr>
<tr></tr>
<tr><td rowspan="2">Main Lines of Action
工作重点</td><td>MLA1:
Strengthening STI (science, technology and innovation) policies, governance and the science-policy-society interface
工作重点 1：
加强科技与创新政策、治理及科学–政策–社会间的互动</td><td>MLA 2:
Building institutional capacities in science and engineering
工作重点 2：
建设科学和工程学方面的机构能力</td><td>MLA 3:
Promoting knowledge and capacity for protecting and sustainably managing the ocean and coasts
工作重点 3：
增进知识和能力，以保护和可持续管理海洋与沿海地区</td><td>MLA 4:
Fostering international science collaboration for earth systems, biodiversity, and disaster risk reduction
工作重点 4：
促进地球系统、生物多样性和降低灾害风险方面的国际科学合作</td></tr>
<tr><td></td><td></td><td>MLA 5:
Strengthening the role of ecological sciences and biosphere reserves
工作重点 5：
强化生态科学和生物圈保护区的作用</td><td>MLA 6:
Strengthening freshwater security
工作重点 6：
加强淡水安全</td></tr>
</table>

(续表)

	Major Programme II 重大计划 II			
Expected Results 预期成果	ER 1: Strengthening STI policies, the science-policy interface, and engagement with society, including vulnerable groups such as SIDs and indigenous peoples (MLA1) 预期成果 1: 加强科技与创新政策、科学与政策的互动以及与社会的联系，包括与小岛屿发展中国家和原住民等弱势群体的联系（工作重点 1）	ER 2: Capacity-building in research and education in the natural sciences enhanced, including through the use of ICTs(information communication technologies) (MLA2) 预期成果 2: 通过信息通信技术等手段，加强自然科学研究和教育方面的能力（工作重点 2）	ER 3: Interdisciplinary engineering research and education for sustainable development advanced and applied (MLA2) 预期成果 3: 推动并应用有利于可持续发展的跨学科工程学研究与教育（工作重点 2）	ER4: Scientific understanding of ocean and coastal processes bolstered and used by Member States to improve the management of the human relationship with the ocean (MLA3) 预期成果 4: 会员国促进和利用对海洋和海岸过程的科学认识，更好地管理人类与海洋之间的关系（工作重点 3）
	ER 5: Risks and impacts of ocean-related hazards reduced, climate change adaptation and mitigation measures taken, and policies for healthy ocean ecosystems developed and implemented by Member States (MLA3) 预期成果 5: 减少海洋危害的风险和影响，采取适应和减缓气候变化的措施，会员国制订和实施促进健康海洋生态系统的政策（工作重点 3）	ER 6: Member States' institutional capacities reinforced to protect and sustainably manage ocean and coastal resources (MLA3) 预期成果 6: 增强会员国的机构能力，以保护和可持续管理海洋和沿海地区资源（工作重点 3）	ER 7: Global cooperation in the ecological and geological sciences expanded (MLA4) 预期成果 7: 扩大生态和地质科学领域的全球合作（工作重点 4）	ER 8: Risk reduction improved, early warning of natural hazards strengthened and disaster preparedness and resilience enhanced (MLA4) 预期成果 8: 进一步降低风险，加强自然灾害的早期预警，提高备灾能力和复原力（工作重点 4）
	ER 9: Use of biosphere reserves as learning places for equitable and sustainable development and for climate change mitigation and adaptation strengthened (MLA5) 预期成果 9: 将生物圈保护区作为促进公平的可持续发展以及减缓和适应气候变化的学习场所（工作重点 5）	ER 10: Responses to local, regional and global water security challenges strengthened (MLA6) 预期成果 10: 加强当地、地区和全球水安全问题的应对措施（工作重点 6）	ER 11: Knowledge, innovation, policies, and human and institutional capacities for water security strengthened through improved international cooperation (MLA6) 预期成果 11: 通过加强国际合作，增强水安全方面的认识、创新、政策及人员和机构能力（工作重点 6）	

IRCK devoted itself to using scientific research results to promote policy making and social service, for example, the research on rocky desertification by IRCK is a great support to related policies on rocky desertification control in China, and the karst carbon cycle research by IRCK promoted the karst carbon sink effects into the formal public documents of China. Moreover, IRCK also made great efforts to carry out serial international cooperative projects with the International Big Scientific Plan on "Resources and Environmental Effects of Global Karst Dynamic Systems" as a flagship, focusing on karst water resources development and water security, karst ecological rehabilitation, karst ecosystem protection, karst geoheritage protection and sustainable development, as well as karst geological disaster early warning and prevention. IRCK's related research responded to Main Line Actions 1, 4, 5, and 6 of 38C/5 well.

中心致力于用科学研究成果推动政策制定及社会服务，如中心开展的石漠化研究对中国石漠化治理政策的影响，岩溶碳循环研究促进中国政府将岩溶碳汇正式纳入政府文件；致力于从地球系统科学角度开展以"全球岩溶动力系统资源环境效应"国际大科学计划为旗舰的系列国际合作项目，开展岩溶区水资源开发与水安全保护研究，岩溶区生态修复研究，推进岩溶生态系统保护，实施岩溶地质遗迹保护与可持续开发，促进岩溶地质灾害预警与防治等，较好地契合了38C/5工作重点1、4、5、6的相应需求。

1.3 UN 2030 Agenda for Sustainable Development
联合国 2030 年可持续发展议程

The 2030 Agenda for Sustainable Development was adopted unanimously in 2015 by all UN Member States at its 70th Session of the General Assembly, and taken into effect on 1 Jan 2016. The Agenda has set 17 Sustainable Development Goals (SDGs) and 169 targets. For recent 6 years, IRCK has made great contributions to SDGs such as Clean Water and Sanitation, Industry, Innovation and Infrastructure, Sustainable Cities and Communities, Climate Action and Life on Land.

《2030 年可持续发展议程》于 2015 年在联合国大会第七十届会议上通过，并于 2016 年 1 月 1 日正式启动。该议程共设有 17 项可持续发展目标和 169 个具体目标。六年来，中心通过业务发展，在清洁饮水和卫生设施，产业、创新和基础设施，可持续城市和社区，气候行动，陆地生物等目标领域做出了突出贡献。

With more focused development and more systematic achievements, IRCK has become an integrated, high-efficient, diversified, and three-dimensional international platform during the second six years.

中心在第二个六年期间的发展更加聚焦，成果更加系统，正在逐步成为一个融合、高效、多元而立体的国际舞台。

Name: Stone Forest Global Geopark, World Natural Heritage Site
Location: Shilin Yi Autonomous County in Kunming, Yunnan
Listed as Global Geopark in 2004
Inscribed by World Natural Heritage List in 2007
Summary: The Stone Forest in Shilin, Yunnan is an outstanding example of pinnacle karst. During the late Paleozoic, the area was a coastal-neritic environment, and thousands of meters of limestone and dolomite were deposited, laying a foundation for the later formation of the pinnacle karst. Tectonic uplift raised the area above sea level, and dissolution by groundwater and surface water along rock fracture finally formed the numerous forms of pinnacle karst seen there today.

名称：石林地质公园、世界自然遗产
所在地点：云南省昆明市石林彝族自治县
列入世界地质公园时间：2004年
列入世界遗产时间：2007年
概述：云南石林地质公园是一个以石林地貌景观为主的岩溶地质公园。晚古生代这里为滨海浅海环境，沉积了上千米的石灰岩、白云岩，为形成本区石林地貌奠定了基础。经受后期地壳运动的抬升作用成为陆地，多期次遭受地下水、地表水沿岩石裂隙进行溶蚀，最后形成了组合类型多样的石林地貌景观。

（本页照片由刘宏提供）／ This photo is from Liu Hong

Chapter 2
Organization and Management

第二章　组织建设与管理

Chapter 2 Organization and Management

From 2016 to 2021, IRCK ensured smooth and efficient operations through three aspects: self-enhancement and improvement, external cooperation to reach win-win benefits, talent cultivation and talent network constructions.

2016~2021年，岩溶中心通过自我提升与完善、对外合作与共赢、人才培养与人才网络架构等三个方面的进展确保运营平稳高效。

2.1 Self-enhancement and Improvement
自我提升与完善

IRCK has accumulated abundant operating experience during the first phase of operation. However, the upmost significant mission for an efficient international platform is continuous self-enhancement and improvement, including improvement of the organization, effective guidance to the operation, and expansion of the international platforms. The self-enhancement and improvement will strengthen the operational capacity and improve the operational effectiveness from the basic structures.

经过第一期运行，中心积累了丰富的运行经验。然而，国际平台的高效运行，首要任务是不断提升与完善自身管理水平，主要包括完善组织机构、有效运行指导以及拓展国际平台。如此才能从基础架构上，增强运行实力，提升运行效率。

2.1.1 Improve the organizational structure to ensure efficient operation
完善组织机构，确保高效运行

The improvement of the organizational structure serves as the basis for an efficient operation. It mainly includes two aspects: a clear organizational structure with well-defined responsibilities; and the statutes, planning, and regulations meeting the development.

完善组织机构是高效运行的基础，主要包括两个方面：一是明确的组织管理机构，权责分明；二是合乎发展要求的章程、规划及管理办法。

2.1.1.1 Organizational Structure
组织结构

The director is responsible for the operation of IRCK under the supervision and guidance of the Governing Board (GB); while the Academic Committee (AC) provides suggestions and guidance to the scientific research of IRCK. IRCK has 6 functional divisions, composed of Secretariat, Financial Division, Domestic Affairs Division, Scientific Research Division, Human Resources and Education Division, and Logistic Support Division; 7 scientific research offices, composed of Karst Dynamics and Global Change Research Office, Karst Resources Research Office, Karst Hydrological Research Office, Karst Environmental Geology Research Office, Karst Engineering and Geohazards Research Office, Karst Ecology and Rocky Desertification Research Office, as well as Karst Regional Geology, Geomorphology and Cave Research Office; and 3 technical supporting sections, composed of the Karst Geophysical Survey Center, the Karst Geo-Data Center, and the Karst Testing Center.

中心实施理事会监督指导下的主任负责制，另设学术委员会指导中心学术研究工作。中心的主要职能部门包括秘书处、财务部、内联部、科研部、人事教育部、后勤保障部；中心的七大业务研究部门包括岩溶动力学与全球变化室、岩溶资源室、岩溶水文地质室、岩溶环境地质室、岩溶工程与灾害地质室、岩溶生态与石漠化室、岩溶区域地质地貌与洞穴室；中心的三大技术支撑部门包括岩溶地质探测技术研究与应用中心、岩溶地质数据处理与应用中心和岩溶地质与资源环境测试中心。

From 2016 to 2021, under the guidance of the GB and AC, IRCK has achieved outstanding achievements in the aspects like international cooperative projects, international exchanges and training, social services, and science popularization and consultation through coordinating the operation of different divisions. IRCK has fulfilled the promise set in the renewal agreement, and performed the set functions sufficiently, realizing an active platform to the world.

2016~2021年，中心在理事会和学术委员会（以下简称"两会"）的指导及各部门协调运作下，在国际项目合作、国际交流与培训、社会服务和科普咨询方面取得了突出的成果，完整履行了续签协定的各项承诺，充分发挥了中心的各项职能，较好体现了中心的平台功能。

第二章　组织建设与管理

```
                    ┌─────────────────┐
                    │ Governing Board │
                    │      理事会      │
                    └────────┬────────┘
                             │
          ┌──────────────────┴──────┐     ┌──────────────────┐
          │      Director           │◄────│ Academic Committee│
          │       主任              │     │     学术委员会     │
          └──────────────────┬──────┘     └──────────────────┘
                             │
   ┌──────────┬──────────┬───┴──────┬──────────┬──────────┐
┌──────┐ ┌─────────┐ ┌────────┐ ┌──────────┐ ┌──────────┐ ┌──────────┐
│Secret│ │Financial│ │Domestic│ │Scientific│ │Human Res.│ │ Logistic │
│ariat │ │Division │ │Affairs │ │ Research │ │and Educa-│ │ Support  │
│秘书处 │ │ 财务部  │ │Division│ │ Division │ │tion Div. │ │ Division │
│      │ │         │ │ 内联部 │ │   科研部 │ │ 人事教育部│ │ 后勤保障部│
└──────┘ └─────────┘ └────────┘ └──────────┘ └──────────┘ └──────────┘
```

Scientific Research Offices 业务研究部门	Technical Supporting Sections 技术支撑部门
Karst Dynamics and Global Change Research Office 岩溶动力学与全球变化室	Karst Geophysical Survey Center 岩溶地质探测技术研究与应用中心
Karst Engineering and Geohazards Research Office 岩溶工程与灾害地质室	Karst Geo-Data Center 岩溶地质数据处理与应用中心
Karst Resources Research Office 岩溶资源室	Karst Testing Center 岩溶地质与资源环境测试中心
Karst Ecology and Rocky Desertification Research Office 岩溶生态与石漠化室	
Karst Hydrological Research Office 岩溶水文地质室	
Karst Regional Geology, Geomorphology and Cave Research Office 岩溶区域地质地貌与洞穴室	
Karst Environmental Geology Research Office 岩溶环境地质室	

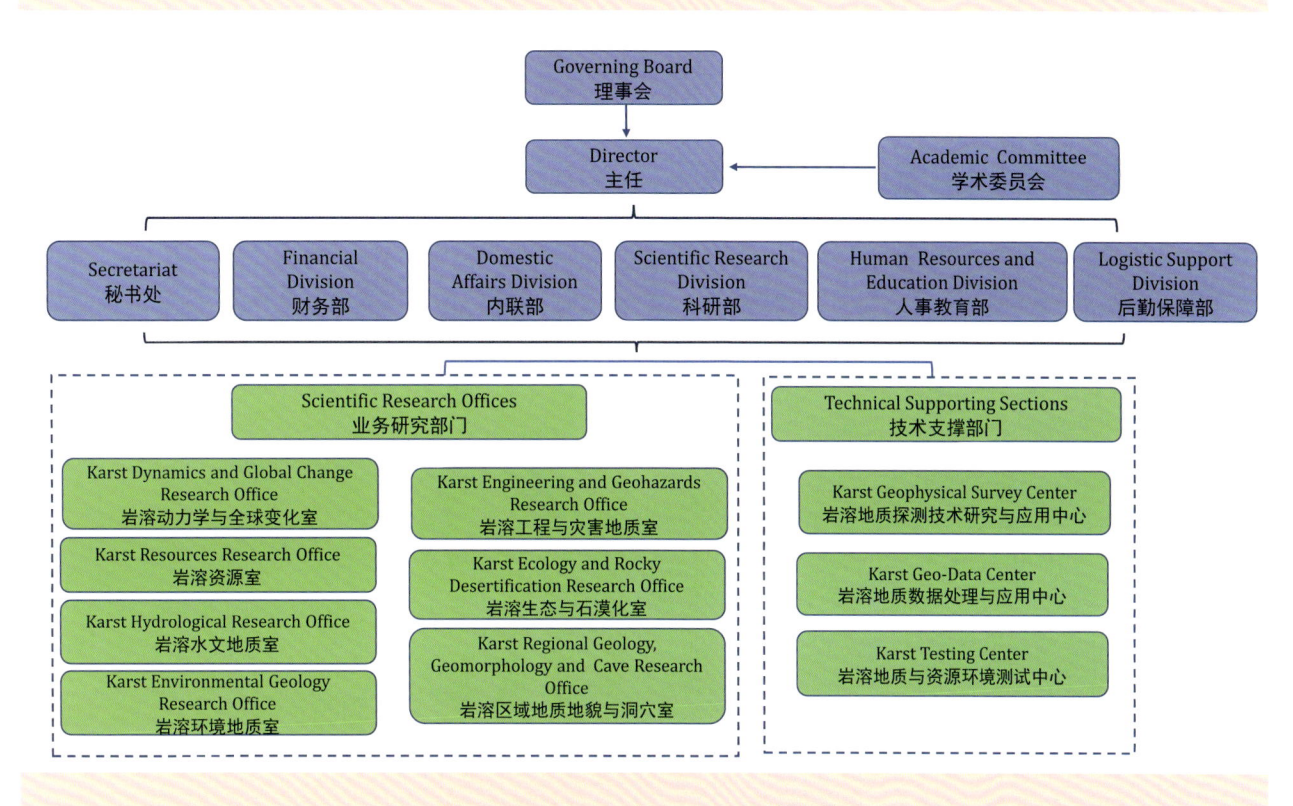

The organizational structure of IRCK
中心组织机构图

2.1.1.2 The Statutes, Plan and Regulations
中心章程、规划及管理办法

The Statutes: IRCK has revised the first-phase statutes, defining more clearly about the requirements to convene the Governing Board Meeting and Academic Committee Meeting, also more clear support from the Government of China.

The Plan: In order to enhance the operational effectiveness and carry out the serial work of IRCK in a planned manner, IRCK has hereby formulated the Six-Year Development Plan for 2016-2021, focusing on capacity building, international exchanges and cooperation, scientific innovation, academic exchanges and training, the publications and outreach, improvement of operational conditions, and the guarantee measures. From 2016 to 2021, IRCK successfully completed serial work defined by the Plan, serving an efficient management mode and sufficient technical support to this international platform.

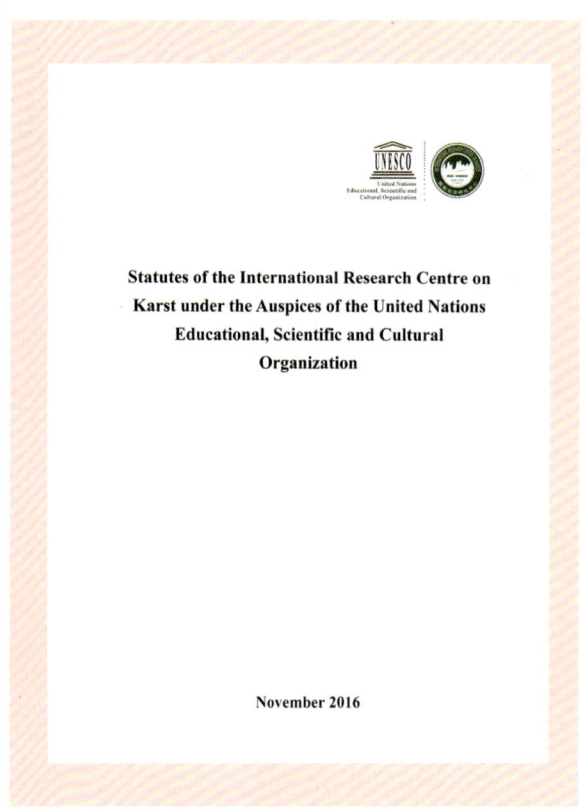

The Regulations: Based on the previous achievements, IRCK revised the Regulations on Finance, Administration, and Personnel, especially, the chapter about Organizational Structure, Personnel, and Duties, which has been revised with a simplified structure by lessening divisions, to reduce the internal management cost and enhance the operational efficiency. The 6-year organic collaboration has verified that the simplified structure enabled more efficient operation.

Statutes of the International Research Centre on Karst under the Auspices of the United Nations Educational, Scientific and Cultural Organization

November 2016

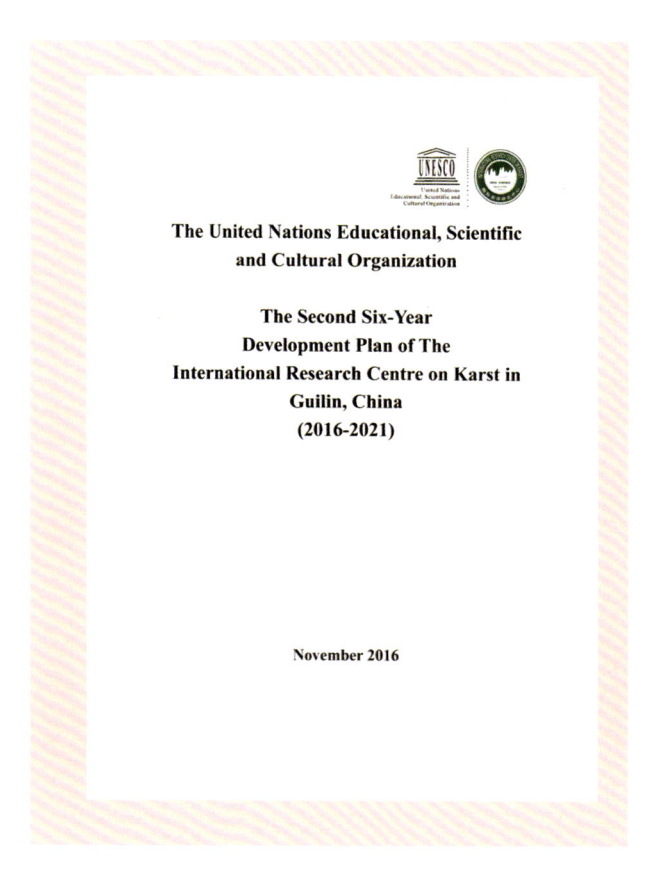

第二章　组织建设与管理

了高效的管理模式与充足的技术保障。

管理办法：在一期运行成果的基础上，中心进一步修改完善了财务、行政和人员管理办法。其中，关于机构、人员与职责部分，相对一期运营进行了机构合并与简化，从而减少内部管理成本，提升运营效率。通过六年的有机协作，证明简化后的机构更加高效。

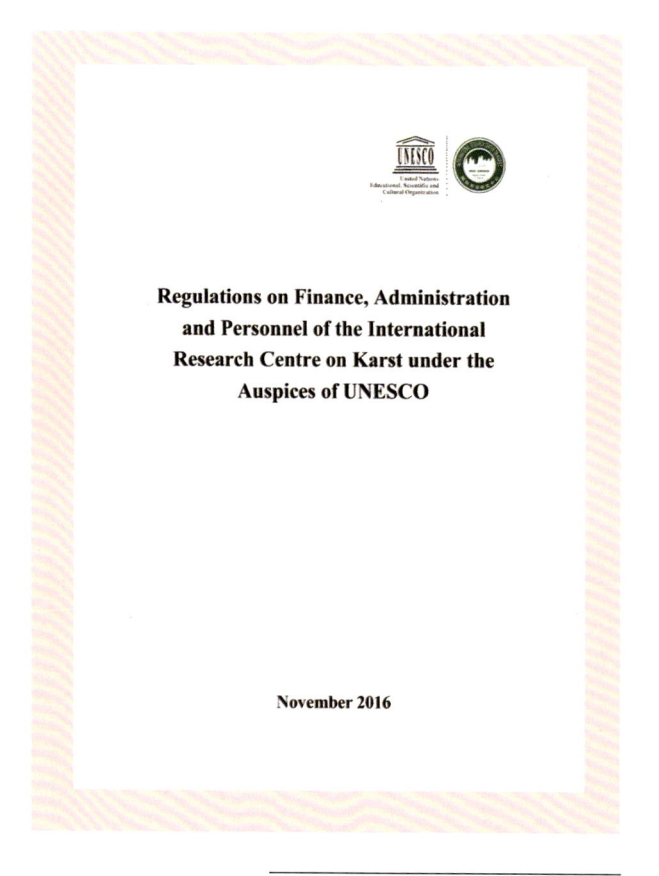

章程：中心修改完善了一期章程相关内容，尤其对理事会和学术委员会的召集规则有了更加明确的要求；我国政府对中心运行的各类支持也得到进一步明确。

规划：为提升中心运营效率，有计划地开展中心系列工作，中心特制定 2016~2021 年二期运营规划。主要围绕如何提升自身能力建设、加强国际交流与合作、加强研究与创新、加强学术交流与培训、加强出版与知识传播、改善中心运营条件以及相应保障措施等方面予以规划。中心较好完成了规划中的系列工作目标，为搭建国际平台提供

The Statutes, Development Plan and Regulations
中心章程、规划及管理办法

国际岩溶研究中心第二个六年历程

2.1.2 Under the guidance of GB and AC, the management reached high quality and high efficiency
两会协同运转，管理质优效高

GB and AC are the two guidance bodies of IRCK. In general, the GB is responsible for the supervision and examination of the daily operation of IRCK, and it acts as the decision maker on the administration, finance, and implementation of the work plans of IRCK; while the AC is responsible for technical advice to the director of IRCK.

理事会和学术委员会是中心的两大指导机构。总体来说，理事会负责中心管理运营的监督与审查，包括中心的行政和财务状况及其工作方案的执行情况；学术委员会则负责向中心主任提供技术领域的咨询意见。

2.1.2.1 The Second Governing Board of IRCK
中心第二届理事会

1. The Second Governing Board Members
中心第二届理事会组成

The Second Governing Board of IRCK has 18 members, including the representatives from the Ministry of Natural Resources of the PRC, UNESCO, International Union of Geological Sciences (IUGS), National Commission of the People's Republic of China for UNESCO (NCC), China Geological Survey (CGS), Chinese Academy of Geological Sciences (CAGS), Department of Science and Technology of Guangxi, Department of Natural Resources of Guangxi, China University of Geosciences (Wuhan), Guilin Bureau of Science and Technology, and the Institute of Karst Geology, CAGS; and the excellent scientists' representatives from the USA, Serbia, Canada and China.

中心第二届理事会共由18人组成，由来自中国自然资源部、联合国教科文组织、国际地质科学联合会、中华人民共和国联合国教科文组织全国委员会、中国地质调查局、中国地质科学院、广西科学技术厅、广西自然资源厅、中国地质大学（武汉）、桂林市科学技术局、中国地质科学院岩溶地质研究所（以下简称"岩溶所"）的机构代表，以及美国、塞尔维亚、加拿大和中国科学家代表等组成。

2. The Second Governing Board sessions of IRCK
中心第二届理事会会议

According to the Statutes, the Second Governing Board (GB-II) sessions convene at least once every two years. From 2016 to 2021, a total of 3 sessions were held in 2016, 2018, and 2020 respectively.

依据中心章程，中心理事会每两年至少召开一次会议。2016~2021年，中心理事会分别于2016年、2018年、2020年共召开了3次会议。

1) The first session of the Second Governing Board
中心第二届理事会第一次会议（2016）

On 14 Nov 2016, the first session of the Second Governing Board (GB-II) was

左图 /Left
Mr. Peng Qiming, chairperson of the Governing Board
理事会主席彭齐鸣先生

中图 /Middle
Mr. Hans Thulstrup, the representative from UNESCO
教科文代表汉斯·图尔斯特鲁普先生

右图 /Right
Mr. Chris Groves, the scientist representative from the USA
美国科学家代表克里斯·葛立夫先生

Chapter 2   Organization and Management

From left to right
Mr. Petar Milanovic, the scientist representative from Serbia; Academician Yuan Daoxian, the scientist representative from China and Academician Wang Yanxin, also the scientist representative from China.

从左至右
塞尔维亚科学家代表皮特·米拉诺维奇先生、中国科学家代表袁道先院士、王焰新院士。

held in Guilin. Then there were 17 Governing Board members, and 15 of them attended the meeting, the participation rate is more than two-thirds. During the meeting, the Governing Board members debriefed the outcomes of 2014–2016, the Statutes, the Six-Year Development Plan (2016–2021), and the Regulations; moreover, the Governing Board debriefed the report on the International Big Scientific Plan on "Resources and Environmental Effects of Global Karst Dynamic Systems" (Global Karst).

2016年11月14日，中心第二届理事会第一次会议在桂林顺利召开。本届理事共由17人组成，出席会议的理事为15人，到会人数超过三分之二。会上，理事们听取了中心2014~2016年度的工作成果，中心章程，中心第二个六年发展规划（2016~2021年），中心财务、行政和人员管理办法；同时，听取了中心起草的"全球岩溶动力系统资源环境效应"国际大科学计划的汇报。

After the enthusiastic discussions, the GB-II members recognized the achievements of lRCK from 2014 to 2016 and approved the Statutes and the Regulations preliminarily. They also agreed to take the Global Karst as the major and key task for the next six years, reminding that the Global Karst should be further improved with more focus on the urgent issues on karst at global scale. All the GB-II members extended their great willingness to support and promote IRCK, and common expectation for a bright future of IRCK.

经过理事们热烈的讨论，初步认可中心 2014~2016 年度工作成果，以及中心章程和中心财务、行政和人员管理办法；理事们初步认可将国际大科学计划作为中心第二个六年发展规划（2016~2021 年）的核心内容；强调国际大科学计划需要进一步完善，集中于全球迫切需要解决的科学问题。中心全体理事表示将全力支持、协助推进中心工作，期待中心未来取得更大成绩。

左图 / Left
Mr. Peng Qiming, the chairperson, granted the certificates to the members: Mr. Peng Qiming granted the certificate to the representative from UNESCO
中心理事会主席彭齐鸣先生给 17 位理事颁发证书：为教科文代表颁发理事证书

右图 / Right
Mr. Peng Qiming granted the certificate to Mr. Cheng Qiuming, then president of IUGS
彭齐鸣先生为国际地质科学联合会时任主席成秋明先生颁发理事证书

The group photo of the first session of GB-II
中心第二届理事会第一次会议合影

2) The second session of the Second Governing Board
中心第二届理事会第二次会议（2018）

On 14 Nov 2018, the second session of the Second Governing Board was held in Nanning, Guangxi Zhuang Autonomous Region. Entrusted by Chairperson Peng Qiming, the GB-II member Mr. Li Jinfa, vice president (now president) of CGS under the Ministry of Natural Resources (MNR) chaired the meeting. During the meeting, the members debriefed the detailed report about the work progress (2017–2018) and the work plan (2019–2020), and spoke highly of the impressive results in basic researches and practical applications. The members gave great comments on the fruitful achievements in 2017–2018; meanwhile, they suggested that IRCK could try to make some breakthroughs in the aspects like the international cooperation along "The Belt and Road", the application for state key laboratory and the organization of important international conferences based on current advantages, hoping to play a more important role as an international platform, also as a significant guidance to the karst research.

2018年11月14日，中心第二届理事会第二次会议在广西南宁顺利召开，受彭齐鸣主席委托，中心理事、自然资源部中国地质调查局党组成员、副局长李金发（现中国地质调查局局长）主持会议。会上，理事们听取了中心 2017~2018 年工作进展以及 2019~2020 年规划的详尽汇报，对中心在基础研究、成果应用等方面取得的工作成果表示高度认可；认为中心 2017~2018 年的运行成果丰硕，成效显著。理事们建议中心未来两年的工作可以立足现有优势，争取在"一带一路"国际合作、国家重点实验室申报、国际重大会议申办等方面取得突破，以期更好地体现中心的平台作用，巩固国际岩溶研究引领性发展的地位。

In his concluding speech, Mr. Li Jinfa put forward six proposals concerning the precise positioning of IRCK to better karst work in the new era and sustain the global importance of karst science: first, to focus more on the comprehensive survey of natural resources; second, to take earth system science as theoretical guidance; third, to keep the scientific innovation as the driving force and strengthen the discipline construction,

From left to right

The comments from Mr. Han Qunli (the representative from UNESCO), Academician Yuan Daoxian (the scientist representative from China), and Mr. Chris Groves (the scientist representative from the USA) said that IRCK should strengthen the cooperation with UNESCO and other international organizations to serve for the national strategy of China and sustainable utilization of resources in karst areas all over the world.

从左至右

中心理事教科文组织代表韩群力先生、中国科学家代表袁道先院士、美国科学家代表克里斯·葛立夫先生在会上发言，指出中心应在良好运行的基础上，加强与教科文组织合作，加强与其他国际组织合作，为国家战略大局服务，为全球岩溶区资源环境可持续利用服务。

talent team construction, scientific platforms construction, and international cooperation; fourth, to extend the research from China to the world, also from surface to the deep; fifth, to change project quantity-based survey into the quantity-quality-ecology based survey; and sixth, to promote benefits reform of karst work.

在总结中，李金发理事就中心如何准确定位，做好新时代的岩溶地质工作，继续发挥国际引领作用，提出六点意见：一是工作内容向自然资源综合调查转变；二是以地球系统科学作为岩溶地质工作的理论指导；三是坚持科技创新作为发展动力，加强学科建设、人才团队建设、科技平台建设和国际合作；四是从国内向国际、从地表向深部拓展岩溶地质工作空间；五是岩溶地质调查方式从过去的以项目数量为主体，向以数量、质量、生态一体转变；六是推进岩溶地质工作的效益变革。

左图 / Left
Conclusion remarks from Mr. Li Jinfa, the entrusted chairperson
李金发代理事长做总结发言

右图 / Right
The group photo of the second session of GB-II
中心第二届理事会第二次会议合影

3) The third session of the Second Governing Board
中心第二届理事会第三次会议（2020）

On 21 December 2020, the third session of the Second Governing Board was held in Guilin by "Virtual + On-spot" pattern. A total of 16 members attended the meeting, they were representatives from UNESCO Division of Ecological and Earth Sciences, the International Union of Geological Sciences (IUGS), National Commission of the People's Republic of China for UNESCO (NCC), China Geological Survey (CGS), Western Kentucky University of the USA, Serbia Chapter of International Association of Hydrogeologists, Chinese Academy of Geological Sciences, China University of Geosciences (Wuhan), Guangxi and Guilin administrations on science and technology, as well as the Institute of Karst Geology (IKG).

2020年12月21日，中心第二届理事会第三次会议在桂林召开，会议采取"线下+线上"方式举行，共邀请了来自联合国教科文组织生态与地球科学部、国际地质科学联合会、中国联合国教科文组织全国委员会、中国地质调查局、美国西肯塔基大学、国际水文地质学家协会塞尔维亚分会、中国地质科学院、中国地质大学（武汉）、广西壮族自治区及桂林市科技主管部门和岩溶所等相关机构的16位国内外理事出席。

At the host of Chairperson Peng Qiming, the members reviewed and debriefed the Biennial Work Report for 2019–2020 and the Biennial Work Plan for 2021–2022 on organization and management, scientific research, international cooperation and training, as well as science popularization. They agreed that IRCK has made outstanding achievements during the last 2 years, but gave important suggestions to the Work Plan: it is suggested to strengthen the cooperation with IUGS and try to make great efforts under the leadership of the new president; to start the evaluation and renewal of IRCK's current agreement as early as possible by smooth communication with relevant agencies (e.g. MNR and NCC); to follow the new five-year scientific and technological development plan of China closely; to make more contributions to poverty alleviation; to focus more on new instruments and technology for geohazards prevention for guaranteeing the major constructions and provide other scientific support to the major development strategy of China (e.g. high-speed railway construction, hydropower construction, and urban green ecological development).

在中心理事长彭齐鸣先生的主持下，全体理事审议了2019~2020年工作成果及2021~2022年工作规划，听取了中心在组织管理、科学研究、国际合作交流培训及科普咨询方面的工作总结与规划。理事们一致认为过去两年中心的工作卓有成效，针对中心2021~2022年工作规划，理事们提出了如下重要建议：加强与国际地质科学联合会（IUGS）的沟通，在IUGS新任主席领导下共同做好附属组织工作；尽早着手准备第二期评估与第三期协定续签工作，加强与主管部委及中国联合国教科文组织全国委员会等有关部门的沟通，顺利完成这两项工作；紧随国家新一期科技创新5年发展规划，注重岩溶科学研究对于脱贫攻坚的贡献，新仪器技术在地质灾害防治领域的应用，以及对重大工程建设的保障，为高铁、水电、城镇绿色生态发展等国家重大发展战略规划提供科技支撑。

第二章　组织建设与管理

上图 / Top
Group photo of on-spot attendees
中心第二届理事会第三次会议线下参会人员合影

下图 / Bottom
Group photo of online attendees (vote)
中心第二届理事会第三次会议线上参会人员合影（表决）

国际岩溶研究中心第二个六年历程

2.1.2.2 The Second Academic Committee
中心第二届学术委员会

1. The Second Academic Committee Members
中心第二届学术委员会组成

The Second Academic Committee (AC-II) is composed of 28 members, attracting top karst scientists at home and abroad. With 18 scientists from China and 10 scientists from karst-related research agencies in Austria, Poland, Brazil, Spain, Russia, the United States of America, Germany, Slovenia, and South Africa, etc., AC-II has provided significant guidance to the research development of IRCK, and supported various academic exchanges and mutual visits firmly, serving as an important driving force for promoting karst science.

中心第二届学术委员会共由 28 人组成，汇集了国内外岩溶科学研究的顶级科学家团队。其中，国内科学家代表 18 人，国外科学家代表 10 人，分别来自奥地利、波兰、巴西、西班牙、俄罗斯、美国、德国、斯洛文尼亚和南非等国的岩溶相关科研院所。学术委员给予了中心科研发展方向性指导，为中心开展各类学术交流、人员互访等提供了坚定支持，是推动岩溶科学发展的重要动力。

2. The three sessions of the Second Academic Committee
中心第二届学术委员会三次会议

1) The first session of the Second Academic Committee (2016)
中心第二届学术委员会第一次会议（2016）

The Second Academic Committee (AC-II) of IRCK convened its first session in Guilin of China from November 15 to 16, 2016. The session included two parts, the seminar on administration and the seminar on academic progress. During the administration seminar, the members debriefed the achievements of IRCK from 2014 to 2016, the Second Six-Year Development Plan (2016–2021), and the presentation on the International Big Scientific Plan on "Resources and Environmental Effects of Global Karst Dynamic Systems" (Global Karst) drafted by IRCK, then they proposed suggestions as follows: it is thought that IRCK could take full advantages of the international platform to enhance the monitoring network, with data from current monitoring stations be collected to achieve big data sharing by collaboration with UNESCO and other agencies. The data could be used to compile karst-related maps at a global scale, and also be used for the research of karst systems on their mechanisms, like karst carbon cycle. Meanwhile, it is suggested that IRCK could upgrade current monitoring instruments with the precondition that there are abundant available funds; it is also suggested to carry out related scientific research on hypogene karst, including the distribution of karst aquifers, the oil and gas reservoirs and their features; in addition, it is proposed to carry out high-resolution paleo-climate reconstruction based on stalagmites records.

2016年11月15~16日,中心第二届学术委员会第一次会议在桂林顺利召开。本次会议由管理研讨及学术研讨两部分组成。会上,委员们听取了中心2014~2016年度的工作成果及中心第二个六年发展规划(2016~2021年)。委员们在听取了中心起草的"全球岩溶动力系统资源环境效应"国际大科学计划的汇报后展开了热烈的讨论,认为中心可以充分发挥平台作用,加强监测网络的建设,充分利用已有的监测站点和已经取得的监测数据,协同教科文等有关机构,共同实现大数据的共享机制,并在有效共享这些数据的基础上,编制全球岩溶相关图件,研究岩溶系统的相关机制,包括岩溶碳循环等;同时,在获得充足资金的资助下开展监测仪器设备的更新换代;中心未来还需要在深部岩溶开展相关科研工作,包括岩溶含水层的分布、深部岩溶油气储层分布及其相关特征;此外,中心还将利用石笋记录开展高精度的古气候重建研究等。

During the academic seminar, 11 experts were invited to give presentations on topics such as global climate change, soil and water management in karst areas of southwest China, carbon sinks and carbon cycles in karst areas, karst collapse in China, and karst groundwater development and utilization in southern China.

学术研讨期间,中心邀请了11名专家做学术报告。围绕全球气候变化、中国西南岩溶区水土治理、岩溶地区碳汇与碳循环、中国岩溶塌陷及南方岩溶地下水开发利用等学术问题展开讨论。

Chapter 2 Organization and Management

上左图 / Top left
Co-director Ralf Benischke
学术委员会副主任拉尔夫·比尼斯尔克

其他图 / Others
AC-II members expressed their willingness to make some contributions to Global Karst based on their expertise and continuous efforts.
全体学术委员表示将基于自身擅长的研究领域，全力支持并推进中心的科研工作，逐步推进完成"全球岩溶动力系统资源环境效应"国际大科学计划，为中心未来发展贡献自身力量。

第二章　组织建设与管理　　035

上左图 /Top left
Director Yuan Daoxian granted certificates to the AC-II member Prof. Andrej Tyc
袁道先主任向学术委员安德烈·泰克教授颁发聘书

上中图 /Top middle
Director Yuan Daoxian granted certificates to the AC-II member, Dr. George Veni
袁道先主任向学术委员乔治·维纳博士颁发聘书

上右图 /Top right
Director Yuan Daoxian granted certificates to the AC-II member, Dr. Jiang Zhongcheng
袁道先主任向蒋忠诚博士颁发聘书

下图 /Bottom
The group photo of the first session of AC-II
中心第二届学术委员会第一次会议合影

国际岩溶研究中心第二个六年历程

2) The second session of the Second Academic Committee (2018)
中心第二届学术委员会第二次会议（2018）

The Second Academic Committee (AC-II) of IRCK convened its second session in Nanning of China on November 14, 2018, which was jointly hosted by Director Yuan Daoxian and Co-director Ralf Benischke. At the beginning, the AC-II members, Mr. Tadej Slabe from Slovenia and Mr. Chris Groves from the USA made wonderful keynote speeches themed on the research progress of classic karst areas and karst critical zones respectively; afterwards, Executive Deputy Director Cao Jianhua and Deputy Secretary-General Luo Qukan introduced the operational progress of IRCK from 2017 to 2018, and the work plan from 2019 to 2020 in details successively. The members spoke highly to the research progress and work plan; meanwhile, they paid great attention to the newly drafted *Karst Map of the World (1:10 million)*, providing suggestions to the compilation methods and data collection approaches in the hope that the map can demonstrate more useful details. Moreover, the members proposed the following suggestions for further internationalization of IRCK: to focus more on the comparison of carbonate and non-carbonate rocks, to consider the compilation of previous training materials, and to promote the international standardization of karst-related results, by taking full advantage of the platform and sharing the outcomes.

2018年11月14日，中心第二届学术委员会第二次会议在广西南宁顺利召开，中心学术委员会主任袁道先院士、副主任拉尔夫·比尼斯尔克先生共同主持本次会议。会议首先邀请学术委员会委员、斯洛文尼亚岩溶学家特德·斯莱布先生和美国岩溶学家克里斯·葛立夫先生分别围绕经典岩溶区研究进展、岩溶关键带研究进展做了精彩的学术报告；随后，中心常务副主任曹建华研究员和中心副秘书长罗劬侃女士分别对中心2017~2018年运营成果以及2019~2020年规划做了详细介绍。全体专家对中心取得的进展和规划予以了高度肯定，同时对中心即将发布的《全球岩溶分布图（1∶1000万）》给予了高度关注，从编制方法、方式、数据采集等多方面给出了多项建议，以期该分布图能展现更成熟、更翔实的成果。此外，专家分别对中心碳酸盐岩和非碳酸盐岩对比研究、历年培训的资料整理、岩溶国际标准化工作推进等给出了意见和建议，建议中心充分发挥平台作用，将成果和经验分享出来，进一步推进中心国际化运作。

第二章　组织建设与管理　　037

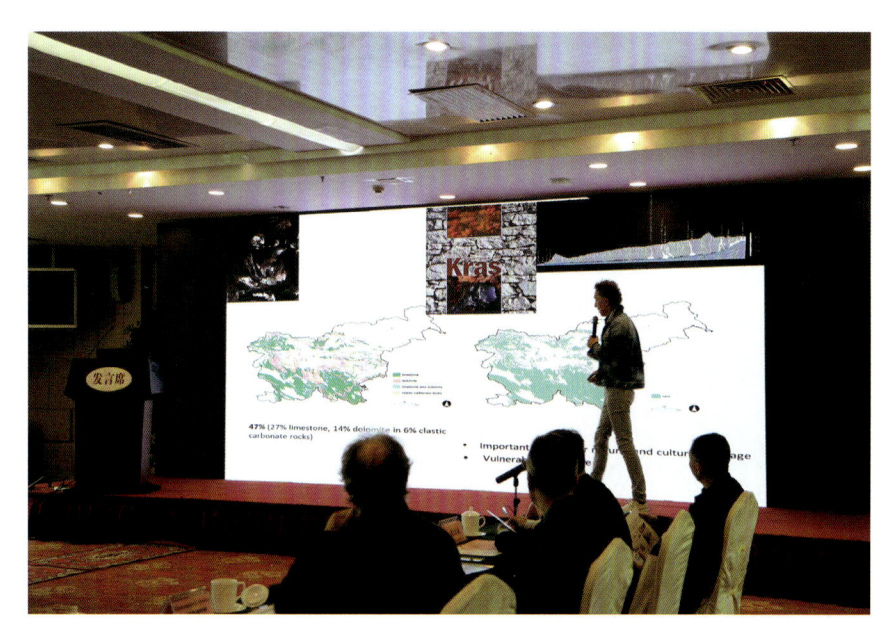

上图 / Top
Mr. Tadej Slabe, director of the Karst Research Institute of Slovenia, made a keynote speech on karstology in the classic karst
斯洛文尼亚岩溶所所长特德·斯莱布先生做经典岩溶区研究进展的报告

下图 / Bottom
Mr. Chris Groves of Western Kentucky University made a keynote speech on karst critical zone
美国西肯塔基大学克里斯·葛立夫先生做岩溶关键带研究进展的报告

国际岩溶研究中心第二个六年历程

Chapter 2 *Organization and Management*

上五图 / Top five

The AC-II members have an in-depth discussion about the *Karst Map of the World (1:10 million)*
中心学术委员针对《全球岩溶分布图（1:1000万）》等研究内容畅所欲言、深入探讨

The group photo of the second session of AC-II

中心第二届学术委员会第二次会议合影

3) The third session of the Second Academic Committee (2021)
中心第二届学术委员会第三次会议（2021）

The Second Academic Committee (AC-II) of IRCK convened its third session virtually on Dec 7, 2021. A total of 14 members from Brazil, Poland, the USA, Serbia, Slovenia, and China attended the session. The members debriefed the report on the operational progress during 2019 to 2021 from 7 aspects, including the organization and management, scientific research, academic exchange, international training, scientific popularization, construction of new IRCK base, and activities for supporting the International Year of Caves and Karst (IYCK); afterwards, they listened to the introduction of the Eight-Year Development Plan (Draft) that covers 4 strategic objectives (disciplinary development, social services, global human resources network, and three-dimensional international platform) and several sub-objectives to support the sustainable development of karst areas.

2021年12月7日，中心在线举办了第二届学术委员会第三次会议。来自巴西、波兰、美国、塞尔维亚、斯洛文尼亚和中国的14名专家学者参与了此次会议。会议听取了中心2019~2021年成果报告，内容涵盖组织管理、科学研究、学术交流、国际培训、科学普及、新基地建设、国际洞穴与岩溶年活动等七个方面，以及未来八年发展规划（草拟稿），规划内容涵盖四个战略目标（学科发展、社会服务、全球人才网络、立体国际平台建设）和若干子目标，以服务和支撑岩溶地区可持续发展。

The members highly recognized the achievement of IRCK in the past 3 years and agreed that IRCK made important contributions to behaving actively as an international platform for international cooperation, academic exchange, and international training on karst even with the bad influence of COVID-19. They paid high attention to the Plan with the following suggestions proposed: first, to strengthen the cooperation with UIS, Commission on Karst Hydrogeology of IAH, and Karst Commission under IGU to promote the implementation of Global Karst; second, to accelerate the steps to set up the big data platform on global karst for sharing; third, to supplement the international case studies to the Technical Specifications on Karst Critical Zone Monitoring for better application over the world; fourth, to continue the support to IYCK for raising the public awareness on karst; fifth, to share the labs and carry out global change research by speleothem jointly; sixth, to improve current surveying technology and methods for prevention and monitoring of possible geohazards during the establishment of major constructions in karst areas; seventh, to organize large-scale international conferences, to promote academic exchange, especially during the sessions of Academic Committee to put karst science going forward commonly.

全体委员充分肯定了中心取得的成绩，一致认为中心克服疫情影响，持续为岩溶领域的国际合作、学术交流、国际培训提供平台支撑，做出了重要的贡献。全体委员对中心未来八年发展规划高度关注，并进一步提出了具体建议：一是加强与国际洞穴联合会、国际水文地质学家协会岩溶专业委员会、国际地理联合会岩溶专业委员会的合作，进一步推动"全球岩溶"国际大科学计划的实施；二是加快建立全球岩溶大数据平台，共享全球岩溶数据；三是丰富和完善ISO/TC 319提出的"岩溶关键带监测技术标准"的国际研究案例，以获得全球通用标准；四是继续参与并提升国际洞穴与岩溶年的影响力，提高公众对岩溶的认知；五是共享实验平台，运用洞穴沉积物联合开展全球变化相关研究；六是改进现有岩溶勘探技术和方法，预防和监测岩溶区重大工程可能产生的地质灾害；七是举办大型国际会议、促进人员交流，尤其是充分利用好学术委员会会议，开展专题学术交流，共同推进岩溶学科进步。

The suggestions will be adopted to improve the Eight-Year Development Plan and implemented step by step for establishing a high-quality and efficient international platform.

中心将在委员建议的基础上进一步完善八年规划，分步实施，为世界岩溶科学发展提供一个高效高质的国际平台。

Chapter 2 *Organization and Management*

Virtual group photo of the third session of AC-II

中心第二届学术委员第三次会议合影

2.1.3 Multiple platforms enable powerful disciplinary development
平台建设成果突出，学科发展动力强劲

During 2016–2021, IRCK has joined two international organizations and promoted to establish one international bilateral cooperation platform successfully that focus on international cooperation and scientific research, driving the disciplinary development of karst powerfully.

2016~2021 年，中心成功加入 2 个国际组织，推进设立 1 个国际平台。围绕国际合作、科学研究等开展高效合作，为岩溶学科发展提供强劲动力。

2.1.3.1 Technical Committee on Karst under the International Organization for Standardization (ISO/TC 319)
国际标准化组织岩溶技术委员会

In 2018, IRCK boosted the application for the Technical Committee on Karst under the International Organization for Standardization (ISO/TC 319) actively, promoted the application through video in April, and gained approval from ISO in June. In September 2019, ISO/TC 319 was formally established in Guilin of China, which is the first international standardization organization on karst globally. ISO/TC 319 is aiming to compile and promote the international standards on karst environment, karst resources, and karst geological survey, including the establishment of general basic standards, technical standards on investigation and evaluation, as well as the standards on sustainable development of karst resources, environmental protection and restoration and geohazards prevention and mitigation. So far, a total of 29 countries joined the Committee, with 8 countries like China, Canada, and Russia as Participating Members (P Members), and 21 countries like France, Germany, the UK, and Japan as Observing Members (O Members).

2018 年，中心积极推进成立国际标准化组织岩溶技术委员会的申报工作，于当年 4 月参加视频推介和答辩，当年 6 月该申请获得国际标准化组织批准；2019 年 9 月，国际标准化组织岩溶

技术委员会在中国桂林顺利挂牌。国际标准化组织岩溶技术委员会是全球岩溶领域第一家国际标准化机构，其目的是致力于开展岩溶环境、岩溶资源、岩溶地质调查领域的国际标准的制订和推广，包括建立通用基础标准、调查和评价技术标准，以及岩溶资源可持续开发、环境保护修复和防灾减灾技术标准。目前共有 29 个国家成员加入该技术委员会中，其中参与国包括中国、加拿大、俄罗斯等 8 个国家，观察国有法国、德国、英国、日本等 21 个国家。

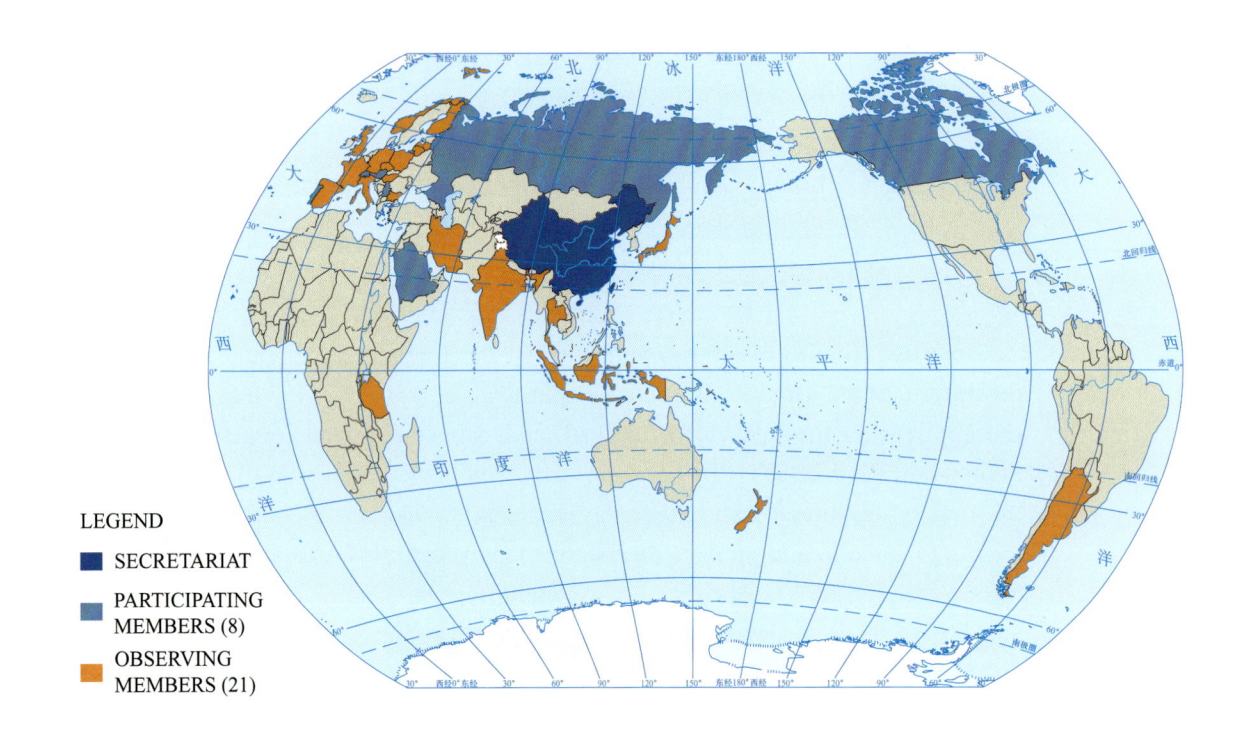

LEGEND
- SECRETARIAT
- PARTICIPATING MEMBERS (8)
- OBSERVING MEMBERS (21)

The Secretariat, Participating Members, and Observing Members of ISO/TC 319

ISO/TC 319 秘书处、参与国与观察国分布图

The list of P Members and O Members of ISO/TC 319
国际标准化组织岩溶技术委员会（ISO/TC 319）参与国与观察国

No. 序号	P Members 参与国	No. 序号	O Members 观察国	No. 序号	O Members 观察国
1	Austria 奥地利	1	Argentina 阿根廷	12	Japan 日本
2	Canada 加拿大	2	Bulgaria 保加利亚	13	Latvia 拉脱维亚
3	China 中国	3	Czech Republic 捷克	14	Lithuania 立陶宛
4	Portugal 葡萄牙	4	Finland 芬兰	15	New Zealand 新西兰
5	Russian Federation 俄罗斯	5	France 法国	16	Norway 挪威
6	Saudi Arabia 沙特阿拉伯	6	Germany 德国	17	Poland 波兰
7	Serbia 塞尔维亚	7	Hungary 匈牙利	18	Spain 西班牙
8	Switzerland 瑞士	8	India 印度	19	United Republic of Tanzania 坦桑尼亚
		9	Indonesia 印度尼西亚	20	Thailand 泰国
		10	Islamic Republic of Iran 伊朗	21	United Kingdom 英国
		11	Italy 意大利		

2.1.3.2 IUGS Affiliated Organization
国际地质科学联合会附属组织

In January 2020, the 74th session of the IUGS Executive Committee approved IRCK for IUGS Affiliated Organization. IUGS has more than 50 affiliated organizations, including the top geological organizations in the world, such as the International Association for Engineering Geology and the Environment (IAEG), the International Association of Geochemistry (IAGC), the International Association of Geomorphologists (IAG), and the International Association of Hydrogeologists (IAH), etc, all of which have corresponding second disciplines for higher education in the field of geosciences. However, karst geology has served as a sub-discipline under hydrogeology for a long time, and the successful application of IRCK as an IUGS Affiliated Organization might

IUGS - International Union of Geological Sciences

Home　What is IUGS?　Organization　Activities　Documents　Publications　Calendar　Contact IUGS

International Association on the Genesis of Ore Deposits (IAGOD)
International Consortium on Landslides (ICL)
International Federation of Palynological Societies (IFPS)
International Geoscience Education Organisation (IGEO)
International Medical Geology Association (IMGA)
International Mineralogical Association (IMA)
International Palaeontological Association (IPA)
International Permafrost Association (IPA)
International Research Center on Karst (IRCK)
International Society for Rock Mechanics (ISRM)
International Society of Soil Mechanics & Geotechnical Engineering (ISSMGE)
National Ground Water Association (NGWA)
The Meteoritical Society (MetSoc)
Society for Environmental Geochemistry and Health (SEGH)
Society for Geology Applied to Mineral Deposits (SGA)
Society for Sedimentary Geology (SEPM)

indicate a possibility to enhance the status of karst geology as important as other second disciplines for higher education. Therefore, in order to realize a better disciplinary status for karst geology, IRCK will continue its cooperative researches, academic exchanges, scientific popularization, and talents cultivation under IUGS.

2020年1月,国际地质科学联合会(以下简称"地科联")召开的第74届执行委员会会议同意将中心纳为国际地科联的附属组织。地科联共有50多个附属组织,囊括了世界顶级地学组织,如国际工程地质与环境协会(IAEG)、国际地球化学协会(IAGC)、国际地貌学家协会(IAG)、国际水文地质学家协会(IAH)等,均为地学领域高等教育二级学科专设国际组织。而岩溶地质长期以来一直是水文地质领域的分支学科,中心成功申报成为国际地科联附属组织,意味着岩溶地质或将与其他二级学科一起占得国际地学领域的一席之位。因此,为推动岩溶地质的学科地位提升,中心将在国际地科联框架下继续开展合作研究、学术交流、科学普及和人才培养等工作。

In 2021, the 76th session of the IUGS Executive Committee approved the annual operational fund for IRCK, which will be mainly used to support the scientists from economic disadvantaged countries or regions to take part in the international workshop and training courses, hoping to enhance the scientific research level of underdeveloped areas.

2021年,国际地科联第76届执行委员会会议批准了中心申请的2021年度运行经费,该经费将主要用于鼓励欠发达地区的科研人员参与中心组织的国际研讨会和国际培训,提升欠发达地区的科研水平,拓展中心影响力。

2.1.3.3 GEO Participating Organization
地球观测组织参与机构

In July of 2021, the Group on Earth Observations (GEO) approved IRCK as a participating organization at its 55th Executive Committee. GEO is a unique global network connecting government institutions, academic and research institutions, data providers, businesses, engineers, scientists, and experts to create innovative solutions to global challenges at a time of exponential data growth, human development, and climate change that transcend national and disciplinary boundaries. The unprecedented global collaboration of experts helps identify gaps and reduce duplication in the areas of sustainable development and sound environmental management. IRCK hopes to join GEO to carry out the cooperation in the fields of sustainable development, climate change, and geohazards mitigation through data sharing, to fill the data blanks for sustainable development of karst areas globally and support the decisions on resources and environmental management.

2021年7月，地球观测组织（GEO）召开的第55届执行局会议通过了中心递交的成为地球观测组织参与机构的申请。地球观测组织是全球唯一的连接政府、科研院所、数据供应商、企业、工程师、科学家和相关专家的国际组织，旨在在数据指数性增长、人类发展和气候变化跨国界、跨学科的时代，提供创新性的解决方案。通过前所未有的专家合作识别可持续发展和健康环境管理领域的空白，减少重复性工作。中心希望加入地球观测组织，通过数据共享开展可持续发展、气候变化、减灾方面的合作，为全球岩溶区社会经济可持续发展填补数据空白，支撑资源环境管理决策。

The approval letter for IRCK as the Participating Organization of GEO
中心获批成为GEO组织参与机构的批准函

2.2 External Cooperation to Reach Win-Win Benefits
对外合作与共赢

2.2.1 Establish closer relationship with UNESCO Headquarters and its field offices to serving science, technology, and innovation development
与教科文总部及其驻外机构紧密联动,助力创新发展

2.2.1.1 Established the mechanism for regular report to UNESCO Headquarters and field offices
建立向教科文总部及教科文驻华代表处定时汇报机制

During the second phase, IRCK has reported regularly to UNESCO Headquarters through the Governing Board sessions (in 2016, 2018 and 2020); moreover, IRCK has reported to UNESCO Headquarters, IGCP Secretariat, and UNESCO Beijing Office through virtual meetings, on-spot visits, and attendance of related meetings held by UNESCO. With the suggestions from different agencies, IRCK has improved its operation and enhanced its efficiency continuously.

二期运营期间,中心除通过理事会会议(2016年、2018年、2020年)定期向教科文总部代表汇报中心进展情况外,还坚持通过线上沟通、实地互访、参与教科文组织相关会议等方式,定期向教科文总部、国际地球科学计划(IGCP)秘书处及教科文驻华代表处汇报中心工作,并在各机构的建议下,不断完善中心运营方式,提升运营效率。

(1) In May 2016, IRCK attended the first UNESCO Science Centres Coordination Meeting. During the meeting, IRCK was selected as the excellent representative of C2Cs and introduced the first-phase operational achievements to all the attendees.

2016年5月,中心参加教科文组织第一届科学领域二类中心主任会议,会上被选为优秀二类中心代表介绍中心一期运营成果。

Chapter 2 Organization and Management

上图 / Top

Executive Deputy Director Cao Jianhua introduced the first-phase operational achievements
中心常务副主任曹建华研究员介绍中心运行情况

下图 / Bottom

The staff of the secretariat took a photo with then UNESCO ADG and specialist on natural sciences from UNESCO Beijing Office
中心秘书处工作人员与教科文时任助理总干事及驻华代表处项目专员合影

The Second 6'Years of IRCK

(2) In February 2017, the 4-member delegation led by Director Liu Tongliang went to UNESCO Headquarters in Paris to report IRCK's progress. The delegation visited the ADG, Ms. Flavia Schlegel, Ambassador Shen Yang of the Permanent Delegation of China in UNESCO, President Cheng Qiuming of IUGS, and Director Han Qunli of the Division of Ecological and Earth Sciences of UNESCO. The in-depth discussion focused on the possible progress of IRCK, and the implementation of the Big Scientific Plan on Global Karst. With abundant achievements, the visits enlightened IRCK greatly for better operation and practical implementation of Global Karst.

上图 / Top
IRCK delegation visited then ADG, Ms. Flavia Schlegel
中心代表团拜访时任教科文组织助理总干事弗莱维娅·斯莱格尔女士

下图 / Bottom
IRCK delegation visited then Ambassador Shen Yang of the Permanent Delegation of China in UNESCO
中心代表团拜访时任中国常驻联合国教科文组织代表团大使沈阳先生

Chapter 2　Organization and Management

上图 / Top

IRCK delegation visited then Director Han Qunli of the Division of Ecological and Earth Sciences of UNESCO
中心代表团拜访时任教科文组织生态与地球科学部主任韩群力先生

下图 / Bottom

IRCK delegation introduced the Big Scientific Plan on Global Karst
中心代表团在教科文总部宣讲"全球岩溶动力系统资源环境效应"国际大科学计划

2017年2月，中心主任刘同良一行4人赴巴黎总部汇报中心工作进展情况。中心代表团拜访了教科文组织助理总干事弗莱维娅·斯莱格尔女士、中国常驻联合国教科文组织代表团沈阳大使、国际地质科学联合会主席成秋明先生，以及教科文组织生态与地球科学部主任韩群力先生。围绕中心工作进展、"全球岩溶"国际大科学计划实施等进行了深入交流与探讨。此访成果丰硕，为提升中心二期运营效率，全面推进"全球岩溶"国际大科学计划提供了重要的启发性和建设性建议。

The Second 6 Years of IRCK

(3) In December 2019, a 3-member delegation led by Director Hu Maoyan visited UNESCO Beijing Office to report the contributions made to SDGs set by the UN *2030 Agenda for Sustainable Development*; meanwhile, both parties expressed their willingness to co-organize an international training course.

2019年12月，中心主任胡茂焱一行3人专程拜访了教科文驻华代表处，汇报了中心为联合国《2030年可持续发展议程》所做的系列贡献，双方还初步表达了共同组织国际培训等系列活动的意愿。

Executive Deputy Director Cao Jianhua made the presentation to the specialist of natural sciences in UNESCO Beijing Office

中心常务副主任曹建华研究员向教科文驻华代表处自然科学项目专员汇报工作

(4) On August 3–4, 2022, Prof. Shahbaz Khan, Director of UNESCO Beijing Office, Representative to China, DPRK, Japan, Mongolia and ROK visited IRCK with his colleague. The visit focused on the second six-year evaluation of IRCK and the management and protection of Guilin World Natural Heritage (WNH) Site. Mr. Peng Xuanming, the director of IRCK, also the director of the Institute of Karst Geology (IKG); Zhao Xiaoming, the deputy director of IKG, and Qin Rongjun, the deputy director of the Lijiang River Scenic Area Administration Committee (LRAC) attended the meeting.

2022年8月3~4日，联合国教科文组织驻华代表处主任、联合国教科文组织驻中国、朝鲜、日本、蒙古和韩国代表夏泽翰教授一行来访中心，针对中心第二个六年评估及桂林世界自然遗产地管理与保护进行了座谈和考察。岩溶所所长、中心主任彭轩明，岩溶所副所长赵小明及桂林漓江风景名胜区管理委员会秦荣军副主任等出席。

Director Peng extended his warm welcome to Prof. Shahbaz Khan and introduced the achievements of IRCK briefly in supporting and serving the UN *2030 Agenda for Sustainable Development* through the following aspects: the effective development and utilization of karst water resources, sustainable development of karst landscape resources, comprehensive treatment of rocky desertification and international training courses on karst since its establishment. Subsequently, Prof. Cao Jianhua, the executive deputy director of IRCK, and Mr. Tang Jianwei, the section director of the LRAC, introduced IRCK's second six-year operation and the Eight-Year Development Plan, as well as the protection and management of Guilin WNH Site. Prof. Khan highly praised the contribution made by IRCK and the LRAC and their support to the sustainable development of karst areas, especially the contributions made by IRCK to the local governments to help to apply for UNESCO Global Geopark and World Natural Heritage, which promoted the leapfrog development of local economy. In view of IRCK's Eight-Year Development Plan, Prof. Khan gave the following suggestions: first, develop and promote global karst research further; second, establish an international karst data platform; third, highlight the international cooperation in the early warning and prevention of geological hazards.

彭轩明主任对夏泽翰教授的来访表示热烈欢迎，并简要介绍了中心成立以来，在岩溶水资源有效开发利用、岩溶景观资源可持续开发、石漠化综合治理及岩溶国际培训等方面支撑和服务联合国《2030年可持续发展议程》的成效。随后，中心曹建华常务副主任和桂林漓江风景名胜区管理委员会（管委会）汤建伟处长分别详细介绍了中心第二个六年运行成果、未来八年发展规划和桂林喀斯特世界自然遗产地保护管理情况。夏泽翰教授对中心及管委会在支撑和服务岩溶区可持续发展方面做出的贡献给予了高度赞扬，对中心在支撑地方政府申请教科文组织世界地质公园和世界自然遗产地，促进地方经济跨越式发展方面给予了充分肯定。针对中心未来八年发展规划，夏泽翰教授给予了以下三点建议：一是大力发展和促进全球岩溶科学研究；二是建立国际岩溶数据平台；三是开展地质灾害预警与防治的国际合作。

After the meeting, Prof. Khan visited the China Museum of Karst Geology (CMKG), IRCK's new base, the experimental site for the research and development of key technologies for sustainable utilization of karst landscape resources in the Lijiang River Basin, and visited the core area of Guilin Karst World Natural Heritage Site. Through this visit, Prof. Khan achieved further understanding of IRCK and the Guilin WNH Site, and

Prof. Khan (row 1, the fourth from the left) visited IRCK and had a meeting with Director Peng Xuanming (row 1, the fourth from the right) of IRCK
夏泽瀚教授（前排左4）一行访问中心并与中心主任彭轩明（前排右4）会谈

emphasized that UNESCO will work together with IRCK and LRAC to support the sustainable development of karst areas.

座谈会结束后，夏泽翰教授先后参观了中国岩溶地质博物馆、国际岩溶研究中心新基地、漓江流域喀斯特景观资源可持续利用关键技术研发与示范工程，并对桂林喀斯特世界自然遗产核心区进行调研。通过为期两天的交流和考察，夏泽翰教授对中心及桂林喀斯特世界自然遗产有了深入的了解，强调教科文组织将同中心、桂林漓江风景名胜区管委会一起共同开展岩溶区可持续发展国际合作。

2.2.1.2 Participated in serial seminars organized by UNESCO field offices to support science, technology, and innovation system development by proposing suggestions
参与教科文驻外机构会议，积极建言献策，助力创新发展

During 2019–2020, IRCK has sent different delegations to attend the virtual or on-spot meetings held by UNESCO field offices, including the Regional Strategic Coordination Meeting "Science Enable and Empower Asia Pacific Strategic for SDGs II" held in Jakarta in 2019 and "Experts Dialogue on SETI Priorities and Implementation Means Asia–Pacific Online Regional Consultation" held in 2020. The meeting focused on the strategic objectives of natural sciences defined by C/5 and SDGs set by the UN *2030 Agenda for Sustainable Development*. IRCK provided suggestions actively, with the hope to enlighten some new ideas in the fields like addressing climate change, promoting earth system science, protecting biodiversity, and reducing the geohazards risks. IRCK would like to get involved in the local communities further to support their economic constructions, such as the supports to Guilin Innovative Demonstration Area for the Sustainable Development Agenda and Guiyang Demonstration Pilot Zone of Ecological Civilization, embracing the 17 SDGs to carry out targeted investigation and cooperation.

2019~2020年度，中心连续派人参与教科文地区机构线下和线上会议，包括2019年度在印度尼西亚雅加达召开的"科学促进亚太地区实现可持续发展目标"——联合国教科文组织区域战略协调会议，以及2020年度在线上召开的"联合国教科文组织亚太区域专家咨询及科学工程技术创新机制"会议。会议围绕教科文自然科学领域C/5中的工作重点以及联合国《2030年可持续发展议程》中的相关目标开展讨论与交流。中心积极建言献策，从岩溶地质角度提供了科技发展创新与国际科学合作的新思路，尤其是在应对气候变化、推动地球系统科学深入研究、保护生物多样性和降低灾害风险方面给出了建设性建议。中心还将深入参与社区经济建设，包括支撑桂林市可持续发展创新示范区建设、贵阳市生态文明试验区建设，围绕17个可持续发展目标，开展有的放矢的调查研究与合作。

The delegates of IRCK, Dr. Cao Jianhua (the second from the right) and Dr. Chen Weihai (the first from the right), together with other delegates from C2Cs in China took group photo with Mr. Hans Thulstrup (the third from the right), the official from UNESCO Jakarta Office.
中心参会代表曹建华副主任（右2）、陈伟海副总工（右1）携其他在华二类中心代表与教科文驻雅加达项目专员汉斯·图尔斯特鲁普先生（右3）合影。

2.2.2 Participated in the serial activities held by the National Commission of the People's Republic of China for UNESCO (NCC) actively, strengthened the collaboration with C2Cs in China and promote the fulfillment of SDGs jointly
积极参与中国全委会系列活动，加强与中国二类中心的沟通协作，协同推进可持续发展目标实现

2.2.2.1 Strengthen the communication with NCC to improve the administration of IRCK
加强与中国全委会的相互沟通，提升中心管理成效

During the second phase, IRCK has strengthened the mutual communication with NCC by visits, invitations, meetings, and submissions of materials in a timely manner to report IRCK's progress, helping NCC to provide in-depth and in-width guidance to IRCK. In addition, IRCK has invited NCC representative to join the Governing Board of IRCK for better management and operation.

二期运行期内，中心加强了与中国联合国教科文组织全国委员会（以下简称"全委会"）的相互沟通。通过拜访、邀请、参会、提交运营材料等方式，及时向全委会汇报中心工作进展，全委会对中心工作的指导更加全面深入。此外，中心通过邀请全委会代表出任中心理事，进一步充实了中心理事会组成，完善了中心管理机构。

1. Visit to NCC
中心一行访问全委会

On 9 September 2020, IRCK delegation visited the Secretariat of NCC, the Secretary-General of NCC extended his warmest welcome to IRCK and his willingness to support IRCK on international cooperation and communications. Currently, there are 15 C2Cs in China. The Government of China pays greater attention to the operation of C2Cs. The Secretariat of NCC spoke highly about the achievements on the construction and application of UGGps, thinking that UGGps not only meet the demands for the poverty alleviation strategy and environmental protection policy of China but also contribute to the SDGs based on the advantages of IRCK. The visit helped IRCK to improve the daily operation and promote kinds of international cooperation conforming to the strategic objectives of UNESCO and the development plan of China.

2020 年 9 月 9 日，中心一行赴全委会秘书处进行汇报交流。全委会秘书长在座谈中首先对中心积极沟通汇报工作情况表示欢迎，支持中心开展符合联合国教科文组织可持续发展目标的国际合作与交流工作。目前，我国已有 15 个二类中心，我国政府对于二类中心的运行更加关注。全委会秘书处对于中心在建设联合国教科文组织世界地质公园领域取得的突出成就表示赞赏，认为既符合中国扶贫战略与环境保护政策，亦有益于联合国可持续发展目标的实现，发挥了中心的业务特色。此访推动中心完善日常管理，推进中心继续开展符合中国发展规划及教科文发展战略的各类国际合作。

The IRCK delegation visited the Secretariat of NCC
中心一行访问全委会秘书处

Attended the meeting of UNESCO Partners in 2019 (left) and 2023 (right)

参加 2019 年（左）和 2023 年（右）中国联合国教科文组织全国委员会合作伙伴座谈会

2. Participating annual meeting held by NCC
积极参加全委会系列年会

IRCK has taken part in serial annual meetings held by NCC actively. On 30 April 2020, NCC convened a virtual meeting on "Driving the Joint Efforts to Fight Against the COVID-19 Pandemic with UNESCO", IRCK shared the administration experience with other C2Cs, Creative Cities, Chairs, and partners. IRCK expressed a strong willingness to contribute more to the fight against the pandemic. On 29 December 2020, IRCK was selected as one of the representatives from the C2Cs in China to share its current achievements and work plan with all the attendees online.

中心积极参加全委会系列年会。2020 年 4 月 30 日，为应对突如其来的全球疫情，全委会召开了"推动与联合国教科文组织合作共抗疫情在线交流会"，中心与其他二类中心、创意城市、教席及合作伙伴等进行了管理经验分享和交流发言，并表达了共抗疫情的强烈意愿，获得了全委会的赞赏。2020 年 12 月 29 日，中心作为二类中心代表之一进行了在线发言，与参会代表分享了中心的工作成效与未来规划。

2.2.2.2 Active coordination with other C2Cs in China to promote the fulfillment of 2030 SDGs jointly
与在华二类中心积极沟通协作，协同推进联合国 2030 年可持续发展目标实现

In April 2015, IRCK joined China UNESCO C2Cs Alliance (the Alliance), the Alliance is aiming to provide a platform to share information among C2Cs in China, to strengthen the collaboration, and to take full advantages of different fields, so as to make some contributions to the strategic objectives of UNESCO. In September 2015, IRCK organized the international training course on "Karst Landscape, Geopark, Natural Heritage Sites, Environmental Geology Mapping and Data Mining", together with UNESCO International Centre on Space Technologies for Natural and Cultural Heritage (HIST) and UNESCO International Knowledge Centre for Engineering Sciences and Technology (IKCEST). This is the first time for IRCK to cooperate with other C2Cs to carry out detailed cooperation, laying an important foundation for the cooperation among C2Cs during the 2nd phase operation.

中心于 2015 年 4 月加入中国联合国教科文组织二类中心联盟，该联盟旨在搭建中国二类中心交流平台，加强相互联动，充分发挥各中心在不同领域的优势，为 UNESCO 战略目标的实现做出贡献。中心曾于 2015 年 9 月，联合国际自然与文化遗产空间技术中心（HIST）和国际工程科技知识中心（IKCEST）等共同举办"岩溶景观、地质公园、自然遗产地、环境地质编图与数据挖掘"的国际培训班。这是中心首次尝试与其他二类中心开展具体合作，为中心二期运营期间开展有效的二类中心合作奠定了重要基础。

In 2016-2021, IRCK has coordinated and cooperated with other C2Cs mainly through the annual meetings and visits, taking full advantages of multidisciplinary absorbing education, science, and culture to put forward the fulfillment of 2030 SDGs.

2016~2021 年，中心主要通过参加联盟召集的年会及实地访问等方式加强与其他二类中心的沟通协作，充分利用教育、科学、文化多领域跨学科的平台优势，共同推进联合国 2030 年可持续发展目标的实现。

IRCK attended the annual Alliance meetings in Beijing (2016), Suzhou (2018), and Taiyuan (2019) respectively. IRCK learned a lot from other C2Cs and proposed suggestions actively, hoping to improve the operation and efficiency continuously by learning and promote the common progress with other C2Cs by sharing.

中心分别在北京（2016年）、苏州（2018年）、太原（2019年）参与二类中心联盟年度会议。会上，中心认真学习其他二类中心成功经验，为推进实现联合国2030年可持续发展目标积极建言献策，通过不断完善自身运营管理机制，提升运营成效，并通过联盟再次分享经验，从而推进在华二类中心共同进步。

On 16 May 2016, Beijing, UNESCO Science Centres Coordination Meeting. Under the instructions of Secretary-General Du Yue of NCC, had discussion on the effective management mechanism and collaboration methods.

2016年5月16日，北京，二类中心联席会议，在全委会秘书长杜越先生的指导下，探讨二类中心有效管理运营的机制与协同合作的方法。

On 29 January 2018, Suzhou, 2018 Alliance Annual Meeting. Had a discussion on the improvement of the Alliance's operational mechanism, exploring new modes for C2Cs' synergetic development.

2018年1月29日，苏州，二类中心联盟2018年年会，探讨进一步完善联盟运行机制、探索二类中心协同发展新模式。

On 21 March 2018, Beijing, Round Table Meeting of C2Cs. Reported the progress made in scientific research, science popularization, and training to Dr. Philippe Pypaert, the newly-appointed specialist on natural sciences of UNESCO Beijing Office.

2018年3月21日，北京，科学类二类中心圆桌会议，向教科文驻华代表处新任自然科学项目专员贝斐然博士汇报各二类中心在研究、科普、培训等方面取得的运行成效和成果。

On 13-15 November 2019, Taiyuan, 2019 Alliance Annual Meeting. Had a discussion on further improvement of the Alliance operational mechanism, exploration of a new mode for coordinated development; meanwhile, learned about the experience of ecological restoration of Xishan Eco-Cultural Tourism Demonstration Zone.

2019年11月13~15日，太原，二类中心联盟年会，探讨完善联盟运行机制、探索二类中心协同发展新模式。此外，学习了解西山生态文化旅游示范区的生态治理情况。

上图 / Top

In March 2018, Dr. Philippe Pypaert, the newly-appointed specialist on natural sciences of UNESCO Beijing Office convened a round table meeting for C2Cs

2018年3月，教科文驻华代表处新任自然科学项目专员贝斐然博士召集在华二类中心召开圆桌会议

下图 / Bottom

In 2019, IRCK attended the 2019 Alliance Annual Meeting

2019年，中心参加在华二类中心联盟2019年年会

IRCK delegation visited IRTCES in September, 2020
2020 年 9 月中心访问国际泥沙研究培训中心

In September 2020, IRCK visited the International Research and Training Center on Erosion and Sedimentation under the auspices of UNESCO (IRTCES), which is the first C2C of UNESCO, also the earliest C2C established in China. During the visit, the two C2Cs introduced their own operations; moreover, IRCK learned a lot of valuable experience from IRTCES like the administration of international associations, SCI journals operation, and the renewal of the agreement; while the IRTCES was enlightened by the application of Technical Committee on Karst under the International Organization for Standardization. The two centers reached a consensus to strengthen the cooperation to improve the operational effectiveness jointly.

2020 年 9 月，中心访问了国际泥沙研究培训中心（泥沙中心），该中心是联合国教科文组织第一个二类中心，也是中国最早的二类中心。此访，双方介绍了各自的运营管理方式，中心在泥沙中心的国际学会组织管理、国际 SCI 期刊管理及二期协定签署上，学习了非常宝贵的经验；泥沙中心则认为中心在国际平台的打造上，尤其是国际标准化组织岩溶技术委员会的申报上对其具有非常重要的启示作用。未来，两个中心将加强合作，共同提升运营成效。

2.3 Human Resources Cultivation and Talents Network Construction
人才培养与人才网络建设

2.3.1 Laid a foundation for talents network by active cultivation
积极实施中青年人才培养计划，奠定岩溶人才网络基础架构

From 2016 to 2021, IRCK has cultivated talents by sending young and middle-aged scientists or technicians to research abroad, introducing international outstanding youths, annual training courses, and cultivating doctoral and master students, among others, laying the foundation for karst-talent network. A total of 8 young and middle-aged scientists or technicians have been to the United States, Germany, and Canada to carry out cooperative research; as of one outstanding international youth has been introduced from Iran; and as many as 139 postgraduates were trained, including 104 masters and 35 doctors. In addition, the 8 training courses organized by IRCK has brought the cumulative trainees benefited to 228 from 44 countries. Mr. Eko Haryono, a trainee from Indonesia, was elected as the chair of the Karst Commission of the International Geographical Union (IGU-KC) in 2018, and Ms. Alena Gessert, a trainee from Slovakia, was elected as Secretary II of European Speleological Federation (FSE) in 2017, both indicating the practical effects of IRCK's training courses.

2016~2021 年，中心通过派出中青年科技人员赴外访学、引进国际杰青、开展年度培训、培养博士和硕士研究生等方式实施中青年人才培养计划，奠定岩溶人才网络基础架构。共派出 8 位中青年科技人员赴外访学，分别赴美国、德国、加拿大等国家开展合作研究工作。引进来自伊朗的国际杰青 1 人。共培养研究生 139 人，其中硕士 104 人，博士 35 人。通过连续举办 8 届国际

岩溶培训班，累计培训学员 228 人次，涉及国别 44 个。来自印度尼西亚的学员艾可·哈约诺先生在 2018 年当选为国际地理联合会岩溶专业委员会主席（IGU-KC），来自斯洛伐克的学员阿琳娜·吉斯特女士在 2017 年当选欧洲洞穴协会秘书长，培训班对人才的培养效果得到了较好体现。

左图 / Left
Mr. Eko Haryono, a trainee from Indonesia, was elected as the chair of the Karst Commission of the International Geographical Union (IGU-KC) in 2018
学员艾可·哈约诺于 2018 年当选为国际地理联合会岩溶专业委员会主席

右图 / Right
Ms. Alena Gessert, a trainee from Slovakia, was elected as the Secretary II of European Speleological Federation (FSE) in 2017
学员阿琳娜·吉斯特于 2017 年当选为欧洲洞穴协会秘书长

2.3.2 Awards to the Governing Board Members
中心理事荣获系列国际奖

2.3.2.1 Serial awards to Mr. Chris Groves (GB-II member) by the government of China
理事克里斯·葛立夫荣获中国政府系列奖励

On the morning of 9 January 2017, the 2016 National Science and Technology Awards Conference was held in the Great Hall of the People in Beijing. On the recommendation of the Ministry of Land and Resources, Prof. Chris Groves, the Governing Board member of IRCK was granted the International Scientific and Technological Cooperation Award of the People's Republic of China, who was received by President Xi Jinping of China. Chris, mainly engaged in the field of karst ecological environment research, is a well-known American expert active in the international karst academic circle. Since 1992, Professor Chris Groves has carried out close cooperation with four organizations (e.g. Institute of Karst Geology, CAGS and Southwest University in China), contributing fabulously to karst geology research in China. In 2019, Chris was awarded the commemorative medal for celebrating the 70th anniversary of the founding of the People's Republic of China. These national-level awards manifested the powerful drive from IRCK to the international cooperation on karst science.

2017年1月9日上午，2016年度国家科学技术奖励大会在北京人民大会堂举行，经国土资源部推荐，中心理事克里斯·葛立夫教授荣获中华人民共和国国际科学技术合作奖，并受到党和国家最高领导人习近平主席的接见。克里斯·葛立夫教授主要从事岩溶生态环境领域的研究，是活跃在国际岩溶学术界的美国知名专家，自1992年以来，克里斯·葛立夫教授与岩溶所、西南大学等4个单位开展密切合作，为中国岩溶地质做出积极贡献。2019年，克里斯·葛立夫先生荣获了庆祝中华人民共和国成立 70 周年纪念章。这些来自国家层面的系列奖励充分说明中心开展的国际合作工作为岩溶科学的发展带来了充足动力。

The certificate and medal for Prof. Chris Groves
克里斯·葛立夫教授所获证书及纪念章

2.3.2.2 Three Governing Board members won Lifetime Achievement Award
中心三位理事荣获终身成就奖

In 2022 June, at its 50th Anniversary of the Karst Commission under the International Association of Hydrogeologists held during EUROKARST 2022, three Governing Board members of IRCK were rewarded the Lifetime Achievement Award, they are Academician Yuan Daoxian from China, Prof. Petar Milanovic from Serbia and Prof. Derek Ford from Canada. Only three scientists were honored to won this award in the world now, which denoted not only the high recognition and respect to them, but also the utmost glory to IRCK. IRCK will continue the exploration and research on the way of karst science, and to find out the exact direction for the long-term development of karst science.

2022年6月，在EUROKARST 2022暨国际水文地质学家协会岩溶专业委员会50周年庆期间，中心三位理事袁道先院士（中国）、皮特·米拉诺维奇教授（塞尔维亚）、戴瑞克·福特教授（加拿大）荣获国际水文地质学家协会岩溶专业委员会终身成就奖，全球仅三位科学家获此殊荣，这是对三位岩溶学家的高度认同与尊敬，也是中心的莫大荣光，中心将继续在三位理事的指导下做好岩溶科学的探索与研究工作，助力全球岩溶科学的长远发展。

Academician Yuan gave suggestions to IRCK (left) and the certificate (right)
袁道先院士为中心建言献策（左）和获奖证书（右）

Chapter 2　Organization and Management

Prof. Petar Milanovic gave suggestions to IRCK (left) and the certificate (right)
皮特·米拉诺维奇教授为中心建言献策（左）和获奖证书（右）

Prof. Derek Ford gave suggestions to IRCK (left) and the certificate (right)
戴瑞克·福特教授为中心建言献策（左）和获奖证书（右）

Name: Wulong World Natural Heritage Site
Location: Chongqing
Inscribed by World Natural Heritage List in 2007
Summary: Wulong World Natural Heritage Site is featured by a carbonate karst landscape, it has combinated and complete geomorphologic features such as caves, sinkholes group, natural bridge group, shaft group, canyons, stone forests, clints, fengcong, fenglin, underground swallet streams, intermittent springs, hot springs, etc. It is a rare karst wonder in China.

名称：武隆世界自然遗产地
所在地点：重庆
列入世界遗产时间：2007年
概述：武隆世界自然遗产地的地质遗迹和地质景观以碳酸盐岩溶地貌为特色，其溶洞群、天坑群、天生桥群、竖井群、峡谷、石林、石芽、峰丛、峰林、地下伏流、间歇泉、温泉等各类地貌分布广泛，组合完好，种类齐全，在全国目前发现的岩溶地貌奇观中实属罕见。

Chapter 3
Scientific Research
第三章 科学研究

Karst covered around 22 million square kilometers globally, accounting for about 15% of the continents, with one quarter population living on karst water (Ford and Williams, 1989). Karst system is a component of earth surface system, with abundant resources, but the fragile ecological environment. With the support of UNESCO and the Ministry of Natural Resources of China, under the common concern from the worldwide karst scientists, IRCK proposed the International Big Scientific Plan on "Resources and Environmental Effects of Global Karst Dynamic Systems" (Global Karst), hoping to realize the sustainable development of society and economy in karst areas.

全球岩溶分布面积约为2200万 km²，约占陆地面积的15%，全球约有1/4人口依赖岩溶水生产生活（Ford and Williams, 1989）。岩溶系统是地球表层系统组成部分，资源丰富，但生态环境脆弱。为实现岩溶区经济社会的可持续发展，国际岩溶研究中心在教科文组织和自然资源部的支持下，在全球岩溶学家的关注下，提出了"全球岩溶动力系统资源环境效应"国际大科学计划（全球岩溶）。

Global Karst, based on the improvement of the karst dynamics theory, aims to promote the sustainable development of global karst areas. It has carried out bilateral and multilateral international cooperation in the fields of karst and climate change (SGD 13), karst water resources and water security (SGD 6), karst rocky decertification and eco-industry (SGD 15), karst geoheritage diversity protection and sustainable utilization (SGD 11), and karst geohazards prevention and early

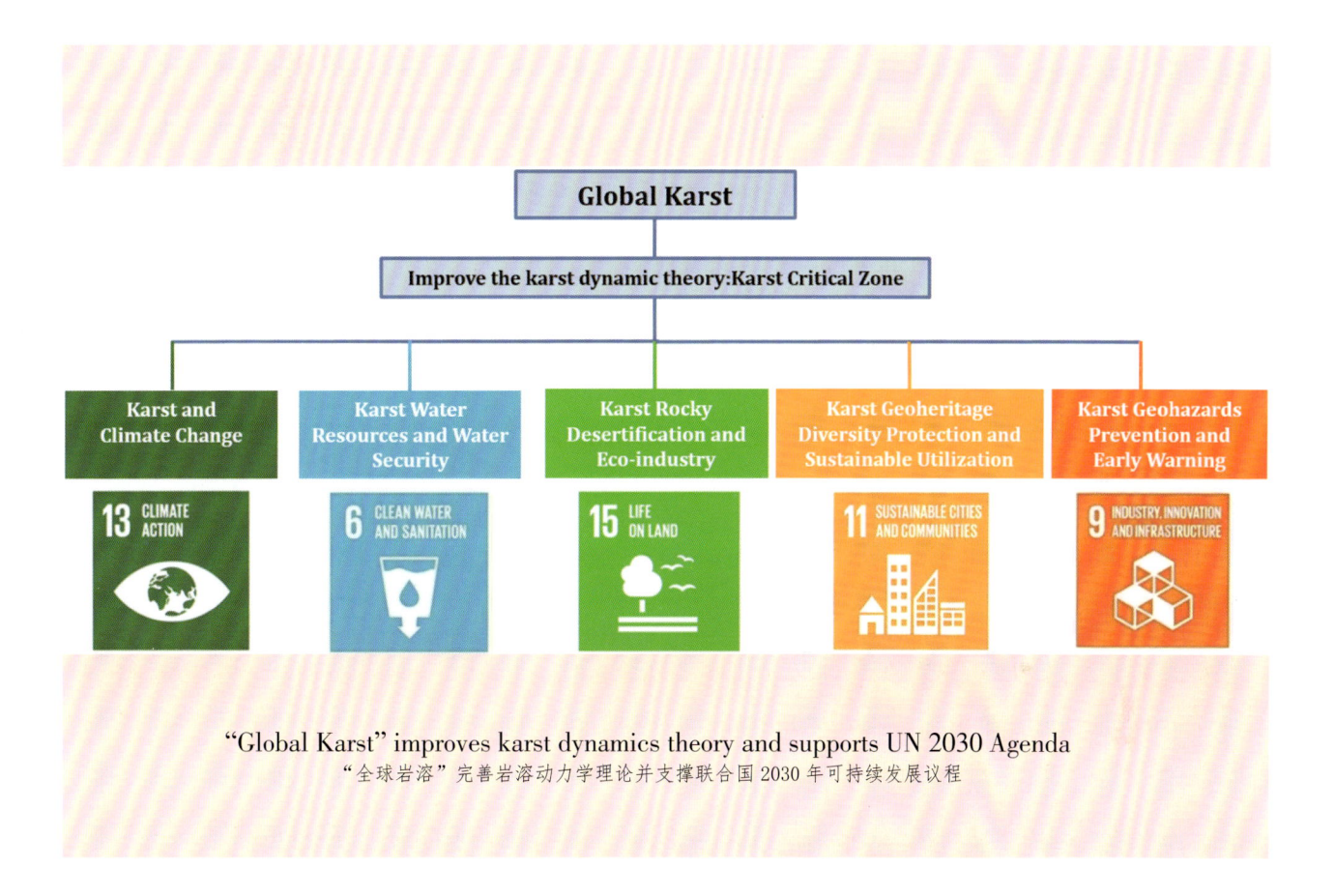

"Global Karst" improves karst dynamics theory and supports UN 2030 Agenda
"全球岩溶"完善岩溶动力学理论并支撑联合国 2030 年可持续发展议程

warning (SGD 9), providing long-term solutions to the global karst resources and environmental problems.

  "全球岩溶"国际大科学计划以促进全球岩溶区可持续发展为宗旨，以完善岩溶动力学理论为基础，围绕岩溶与气候变化（SGD 13）、岩溶水资源与水安全（SGD 6）、岩溶石漠化与生态产业（SGD 15）、岩溶地质遗迹多样性保护与可持续利用（SGD 11）、岩溶地质灾害防治与早期预警（SGD 9）等领域开展多双边国际合作，旨在为全球岩溶区资源环境问题提供长效解决方案。

3.1 International Geoscience Program (IGCP) 国际地球科学计划

In March of 2017, the IGCP 661 entitled as "Processes, Cycle, and Sustainability of the Critical Zone in Karst Systems" was approved and supported by UNESCO and IUGS jointly, with the duration of 2017–2021. The project is a kind of important support from UNESCO to "Global Karst".

2017年3月，国际地球科学计划项目(IGCP 661)"岩溶关键带物质能量循环过程及可持续性研究"顺利获批，该项目得到了联合国教科文组织和国际地科联的共同支持，实施年限为2017~2021年。该项目是教科文组织对"全球岩溶"大科学计划的一项重要支持。

Prof. Jiang Zhongcheng, the GB-II member of IRCK, acted as the leader of the international working group, with 9 scientists from China, the US, Brazil, Slovenia, Iran, Spain and Slovakia as the co-leaders. Moreover, over 70 scientists from Thailand, Russia, Austria and other 23 countries participated in this project.

该项目由中心理事蒋忠诚担任国际工作组主席，来自中国、美国、巴西、斯洛文尼亚、伊朗、西班牙、斯洛伐克的9名科学家担任联合主席，另有来自泰国、俄罗斯、奥地利等26个国家的70余名科学家参与其中。

During the implementation, the international working group held 5 workshops in Kunming of China, Vancouver of Canada, Busan of Republic of Korea, and Guilin of China. Scientists from different countries exchanged progress on hot issues of karst critical zones, such as water cycle and soil leakage; and exchanged the research experience on carbon-water-calcium cycles based on the field work of Maocun and Yaji field experimental sites in the Lijiang River Basin in China. With the implementation of the project, the scientists from internationally renowned universities such as University of Vienna in Austria, St. Petersburg University of Technology in Russia, Kangwon National University in the Republic of Korea, and University of Szczecin in Poland sent letters to IRCK to express their willingness to join the research team of IGCP 661 project, enlarging the international cooperative team.

项目执行期间，国际工作组在中国昆明、加拿大温哥华、韩国釜山、中国桂林召开

第三章　科学研究　077

Leaders and co-leaders of IGCP 661 international working group
IGCP 661 国际工作组主席及联合主席
上图从左至右 / Top row from left to right
Jiang Zhongcheng (China), Yuan Daoxian (China), Zhang Cheng (China), Martin Knez (Slovenia), Chris Groves (the USA)
蒋忠诚（中国）、袁道先（中国）、章程（中国）、马丁·内兹（斯洛文尼亚）、克里斯·葛立夫（美国）

下图从左至右 / Bottom row from left to right
Alena Gessert (Slovakia), Augusto Auler (Brazil), Jiang Yongjun (China), Bartolome Andero Navarro (Spain), Ezzat Raeisi (Iran)
阿琳娜·吉斯特（斯洛伐克）、奥古斯托·奥乐（巴西）、蒋勇军（中国）、巴特洛米·安德鲁·纳瓦罗（西班牙）、伊扎特·瑞斯（伊朗）

国际岩溶研究中心第二个六年历程

了5次工作会议，多国科学家就水循环、土壤漏失等岩溶关键带研究热点问题进行了交流。通过对我国漓江流域毛村、丫吉等野外试验基地进行实地野外考察，交流了岩溶关键带碳－水－钙循环领域的经验方法。随着项目开展，奥地利维也纳大学、俄罗斯圣彼得堡理工大学、韩国江原大学、波兰什切青大学等国际知名高校学者来信表达了加入 IGCP 661 项目研究团队的意愿，国际合作队伍进一步扩大。

Based on the ecosystem of the critical zone restricted by the karst environment, this study analyzed the structural characteristics of the karst critical zone in details: the surface and subsurface are interconnected; the epikarst zone develops intensively, and responses to the external environment rapidly and sensitively; and the soluble rocks are dissolved differentially. It has set up 5 monitoring stations for typical karst

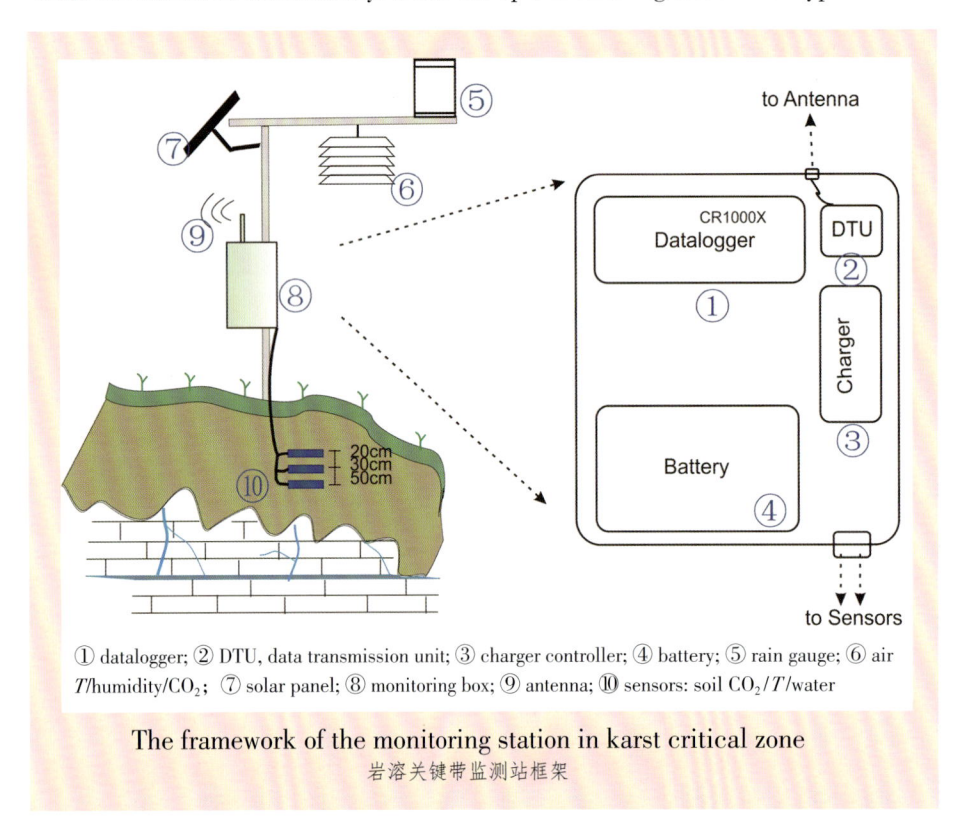

① datalogger; ② DTU, data transmission unit; ③ charger controller; ④ battery; ⑤ rain gauge; ⑥ air T/humidity/CO_2; ⑦ solar panel; ⑧ monitoring box; ⑨ antenna; ⑩ sensors: soil CO_2/T/water

The framework of the monitoring station in karst critical zone
岩溶关键带监测站框架

critical zones, and published more than 200 related papers, enriched the global karst database, and compiled the *Karst Map of the World* (digital product), which promoted the comparative study of global karst critical zones. All the related achievements were concluded in newsletters (Jiang et al., 2019b; Bai et al., 2023).

项目研究以岩溶环境制约关键带生态系统为基础，详细分析了岩溶关键带结构特点：地表与地下相互连通、表层岩溶带发育强烈、迅速而敏感地对外界环境做出响应及可溶岩差异性溶蚀等。建立典型岩溶关键带监测站5个，发表岩溶关键带研究学术论文200多篇，丰富了全球岩溶数据库，编制了《全球岩溶分布图》（数字化成果），推动了全球岩溶关键带对比研究，并以年报形式进行成果汇总（Jiang et al., 2019b; Bai et al., 2023）。

IGCP 661 newsletter (2017–2020)
IGCP 661 年报（2017~2020 年）

3.2 Resarch Achievements
研究进展

3.2.1 Carbon cycle and climate change
碳循环与气候变化

3.2.1.1 Carbon cycle and carbon sink
碳循环与碳汇效应

1. Karst dynamic system and the carbon cycle
岩溶动力系统与碳循环

Karst dynamic system operation mainly includes carbonate rock formation and its dissolution and weathering. The carbonate rock formation made great contribution to atmospheric CO_2 sink in geological history. The carbon stored in carbonate rocks, which amounts to 61×10^{15} t, is known as the largest carbon pool on modern earth, representing 99.55% of the global carbon storage. In modern karst dynamic system, the evidences show the carbonate rock dissolution and weathering actively involves in the global carbon cycle, and sensitively responds to climate and environmental changes, the annual carbon sink flux derived from carbonate rock dissolution amounts to 0.36–0.44 Pg C/a, which is close to the figure of 0.477 Pg C/a, accounting for 32.73%–40.00% of global forest carbon sink of 1.1 Pg C/a from 1990 to 2007. This figure equals 45.00%–55.00% of net carbon flux for global soil organic carbon pool appropriately. The conceptual model of Karst Carbon Cycle in Watershed consists of three parts: carbonate rocks dissolution removes the atmospheric/soil CO_2 to water, then produces inorganic carbon; inorganic carbon transfers and converts along with water flow; aquatic plants photosynthesis enables inter-conversion between inorganic carbon and organic carbon, and part of the organic carbon deposits on the river/lake/reservoir beds by mixing with the sediments (Cao et al., 2016).

岩溶动力系统运行主要包括碳酸盐岩的形成及其溶蚀和风化作用。在地质历史时期，碳酸盐岩的形成对大气 CO_2 汇做出了巨大贡献。碳酸盐岩中的碳储量为 $61×10^{15}$ t，是现代地球上最大的碳库，占全球碳储量的 99.55%。在现代岩溶动力系统中，有证据表明碳酸盐岩溶解和风化积极参与了全球碳循环，对气候和环境变化反应灵敏，碳酸盐岩溶解产生的年碳汇量为 0.36~0.44 Pg C/a，接近 0.477 Pg C/a 的数值。1990~2007 年，岩溶碳汇占全球森林碳汇 1.1 Pg C/a 的 32.73%~40.00%，约等于全球土壤有机碳库净碳流量的 45.00%~55.00%。流域岩溶碳循环的概念模型包括三个部分：碳酸盐岩溶解将大气/土壤中的 CO_2 移入水中，并产生无机碳；无机碳随着水流运移转化；水生植物的光合作用使无机碳和有机碳相互转化，部分有机碳沉积在河流/湖泊/水库底部与沉积物混合（Cao et al., 2016）。

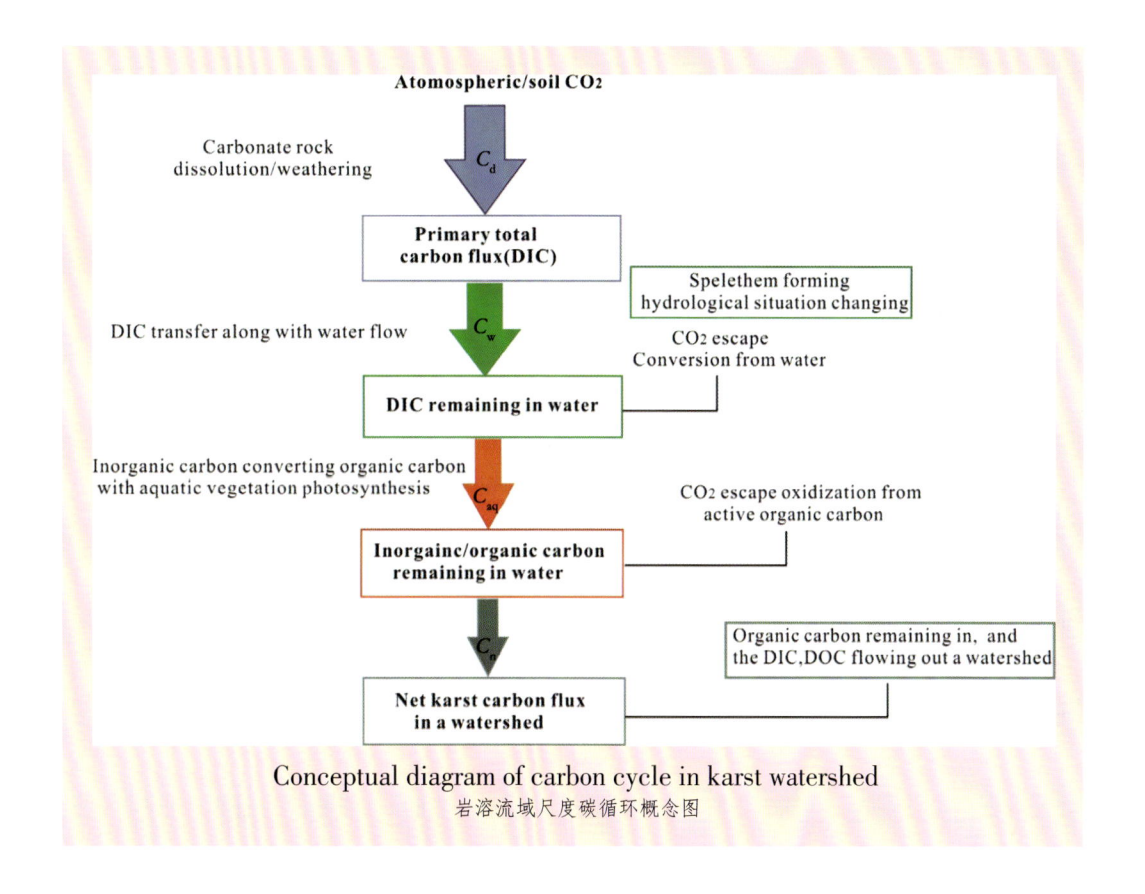

Conceptual diagram of carbon cycle in karst watershed
岩溶流域尺度碳循环概念图

2. Karst process has a short-time carbon cycle effect, which changes the knowledge that geological process does not contribute to modern global carbon cycle

岩溶作用具有短时间尺度碳循环效应，改变了地质作用对现代全球碳循环影响的认知

The inorganic carbon flux of the Pearl River Basin into the sea (calculated by CO_2) is 623×10^4 t/a, with that resulted from the carbonate rocks contributing 80.46%; the organic carbon flux of this basin into the sea (calculated by CO_2) is 608×10^4 t/a, with the endogenous organic carbon output flux as 321×10^4 t/a, and the external organic carbon output flux as 287×10^4 t/a, and the endogenous carbon flux accounts for 52.8%. The total organic carbon deposition in the river (calculated by CO_2) is 4.65×10^4 t/a, with the endogenous carbon deposition as 3.26×10^4 t/a, and the external carbon deposition as 1.39×10^4 t/a, and the endogenous accounts for 70%. The conversion amount of aquatic organisms (calculated by CO_2) is 364×10^4 t/a, accounting for approximately 31% of the inorganic carbon flux in the water body, with 3.26×10^4 t/a entering into the sediments, and additional 321×10^4 t/a entering into the ocean with water flow. The average carbon flux output by river respiration (calculated by CO_2) is 178×10^4 t/a. The net carbon sink in the Pearl River Basin is the sum of CO_2 deposition and sediment input from the atmosphere and the flux transported to the ocean, that is, 949×10^4 t/a, equivalent to 4% of the annual CO_2 emissions of the Pearl River Basin (2.3×10^8 t), which is an important way for carbon emission reduction in this region (Cao et al., 2023).

珠江流域无机碳入海通量（以 CO_2 计）为 623×10^4 t/a，碳酸盐岩贡献率为 80.46%；流域有机碳入海通量（以 CO_2 计）为 608×10^4 t/a，其中内源有机碳输出通量为 321×10^4 t/a，外源有机碳输出通量为 287×10^4 t/a，内源占 52.8%。河流总的有机碳沉积量（以 CO_2 计）为 4.65×10^4 t/a，其中内源碳沉积量为 3.26×10^4 t/a，外源碳沉积量为 1.39×10^4 t/a，内源占 70%。水生生物转化量（以 CO_2 计）为 364×10^4 t/a，约占水体无机碳量的 31%，有 3.26×10^4 t/a 进入底泥中，另外有 321×10^4 t/a 随水流进入海洋。河流呼吸作用平均输出的碳通量（以 CO_2 计）为 178×10^4 t/a。珠江流域净碳汇量为大气输入的 CO_2

沉积到底泥及输送入海的通量之和，即为 949×10⁴ t/a，相当于珠江流域每年 CO_2 排放量（2.3×10⁸ t）的 4%，是该区重要的碳减排项（曹建华等，2023）。

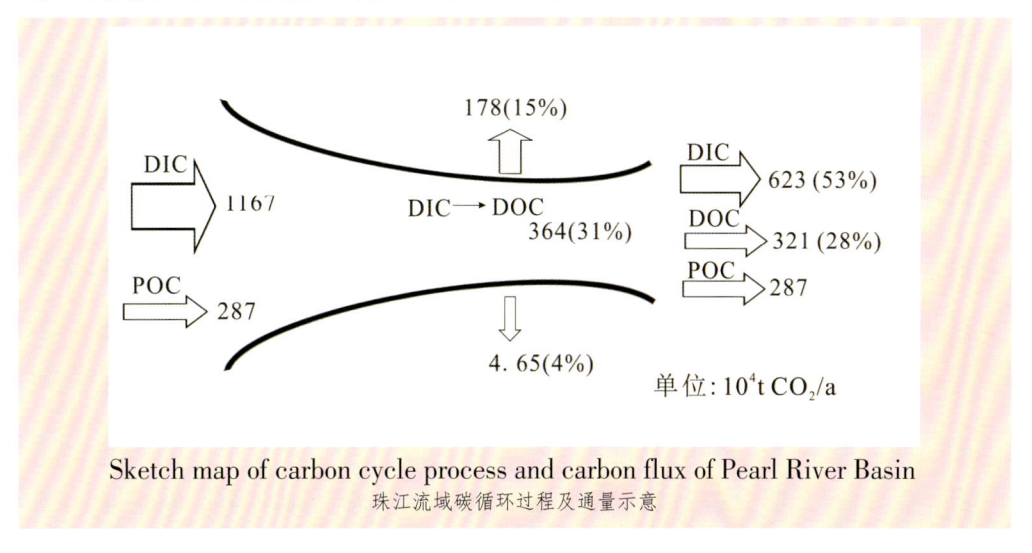

Sketch map of carbon cycle process and carbon flux of Pearl River Basin
珠江流域碳循环过程及通量示意

3. Stable carbon isotopic composition of submerged plants living in karst water and its eco-environmental importance
岩溶水中沉水植物的稳定碳同位素组成及其生态环境意义

The stable carbon isotopic composition of submerged plants ($\delta^{13}C_P$) can be controlled by physiological and environmental factors. Herein, the study took advantage of a short, natural karst river with an annual mean bicarbonate (HCO_3^-) value of 3.8 mmol/L to study the stable carbon isotopic composition of submerged plants along the river and the influence of environmental conditions on the $\delta^{13}C_P$ values. The $\delta^{13}C_P$ values of *Ottelia acuminata*, *Potamogeton wrightii*, *Vallisneria natans*, and *Hydrilla verticillata* from upstream to downstream show a gradient distribution and ranged from −34.78‰ to −27.83‰, −36.56‰ to −23.70‰, −35.06‰ to −25.29‰, and −38.56‰ to −26.32‰, respectively and even more depleted values for the first two species at the uppermost site. Diurnal variation of water chemistry and concentration of the dissolved inorganic carbon (DIC) and the stable carbon isotopic composition of DIC ($\delta^{13}C_D$) indicate that the river has a very high net photosynthetic rate. The gradient distribution of $\delta^{13}C_P$ values was consistent with CO_2 being a declining source of inorganic carbon for photosynthesis in the downstream transect. The results demonstrate that the high DIC concentration with a lower negative $\delta^{13}C$ value, particularly in the karst water environment has a significant role in controlling the stable carbon isotopic composition of submerged plants living in it (Wang et al., 2017).

沉水植物的稳定碳同位素组成（$\delta^{13}C_P$）受控于生理因素和环境因素。在此，本研究以一条较短的自然河为研究对象，根据其年均重碳酸根离子浓度（3.8 mmol/L）来研究沿河沉水植物稳定碳同位素组成及环境因素对$\delta^{13}C_P$值的影响。海菜花（*Ottelia acuminata*）、竹叶眼子菜（*Potamogeton wrightii*）、苦草（*Vallisneria natans*）和黑藻（*Hydrilla verticillata*）的$\delta^{13}C_P$值从上游到下游呈梯度分布，范围分别为 –34.78‰到 –27.83‰、–36.56‰到 –23.70‰、–35.06‰到 –25.29‰和 –38.56‰到 –26.32‰，前两个物种在水体最上层，其$\delta^{13}C_P$值更加亏损。水化学和溶解无机碳（DIC）浓度以及稳定碳同位素组成（$\delta^{13}C_D$）的日变化表明，该河具有很高的净光合速率。$\delta^{13}C_P$值的梯度分布与下游参与光合作用的CO_2无机碳来源减少相一

$\delta^{13}C$ value characteristics of four submerged plants
四种沉水植物的 $\delta^{13}C$ 特征

Species	A	B	C	D	E	F
Ottelia acuminata/‰	–40.48	–34.78	–34.95	–30.61	–34.80	–27.83
Vallisneria natans/‰	–	–35.06	–34.82	–33.32	–34.92	–25.29
Potamogeton wrightii/‰	–39.16	–36.56	–35.83	–33.09	–34.40	–23.70
Hydrilla verticillata/‰	–	–38.56	–	–34.55	–34.00	–26.32
Distance/m	0	85	128	260	332	512

致。结果表明,高 DIC 浓度和较低 $\delta^{13}C$,对控制生活在其中的沉水植物的稳定碳同位素组成具有重要作用,特别是在岩溶水体中(Wang et al.,2017)。

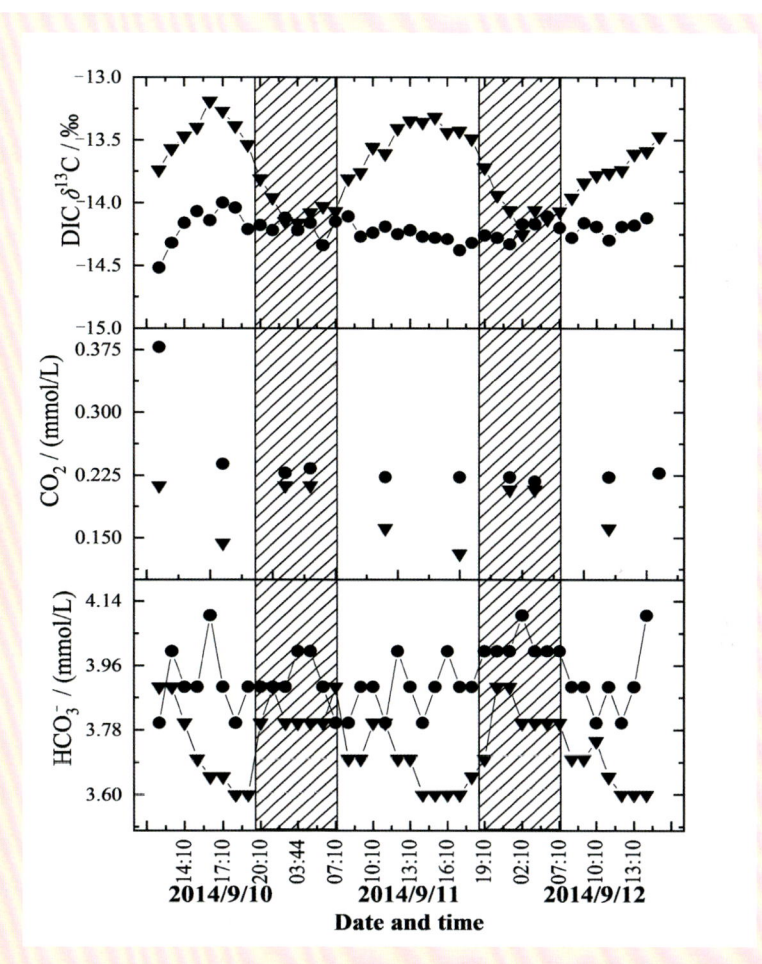

Diel variation of concentration of HCO_3^- and CO_2, and of the $\delta^{13}C$ value of dissolved inorganic carbon at the upstream (circle) and downstream (triangle) sites (Nighttime is shown by hatched shading)

上游(圆圈)和下游(三角)监测点 HCO_3^- 和 CO_2 浓度昼夜变化曲线及溶解无机碳 $\delta^{13}C$ 值昼夜变化曲线(夜间时段以斜线阴影表示)

4. Dynamics in riverine inorganic and organic carbon based on carbonate weathering coupled with aquatic photosynthesis in a karst catchment, Southwest China

中国西南岩溶流域基于水生植物光合作用与碳酸盐岩风化耦合的河流无机碳与有机碳动态变化研究

Carbonate mineral weathering coupled with aquatic photosynthesis, herein termed "coupled carbonate weathering" (CCW), representing a significant carbon sink which is determined by riverine hydrochemical variations. The magnitudes, variations, and mechanisms responsible for the carbon sink produced by CCW are still unclear. In this study, major ions, TOC, and discharge data at the Darongjiang, Lingqu, Guilin, and Yangshuo hydrologic stations in the Lijiang River Basin, a karst catchment typical of this geographic region, were analyzed from January 2012 to December 2015 to elucidate the temporal variations in riverine inorganic and organic carbon and their controlling mechanisms. The results show that ① HCO_3^- was sourced from carbonate weathering and silicate weathering, carbonate weathering by carbonic acid being predominant; ② TOC was created chiefly by the transformation of bicarbonate to organic carbon by aquatic phototrophs during the non-flood period; ③ The carbon sink produced by coupled carbonate weathering in the Lijiang River basin was calculated to be 14.41 t C/(km^2·a), comprised of the sink attributable to carbonate weathering (12.17 t C/(km^2·a)) and sink due to the "biological carbon pump" (S_{BCP}) (2.24 t C/(km^2·a)). The S_{BCP} thus accounted for approximately 15.54% of the total carbon sink, indicating that the proportion of riverine TOC sourced by the transformation from bicarbonate to organic carbon by aquatic phototrophs may be high and must be considered in the estimation of carbonate weathering-related carbon sinks elsewhere (Sun et al., 2021).

碳酸盐矿物风化与水生光合作用耦合（"耦合碳酸盐岩风化"（CCW）），代表由河流水化学变化决定的重要碳汇。耦合碳酸盐岩风化产生碳汇的幅度、变化和机制尚不清楚。本研究分析了 2012 年 1 月至 2015 年 12 月漓江流域典型岩溶流域大溶江、灵渠、桂林和阳朔水文站的主要离子、TOC 和流量数据，以阐明河流无机碳和有机碳的时间变化及其控制机制。结果表明：① HCO_3^- 来源于碳酸盐风化和硅酸盐风化，主要来源于碳酸风化碳酸盐岩；② TOC 主要是通过水生光合生物在非洪水期将重碳酸根离子转化为有机碳而产生；③ 漓江流域耦合碳酸盐岩风化产生的碳汇为 14.41 t C/(km²·a)，包括碳酸盐岩风化产生的碳汇（12.17 t C/(km²·a)）和"生物碳泵"（S_{BCP}）产生的碳汇（2.24 t C/(km²·a)）。因此，生物碳泵约占总碳汇的 15.54%，这表明河流有机碳中，水生光合生物光合利用重碳酸根离子产生的有机碳比例可能很高，须在估算碳酸盐风化产生的碳汇时予以充分考虑（Sun et al., 2021）。

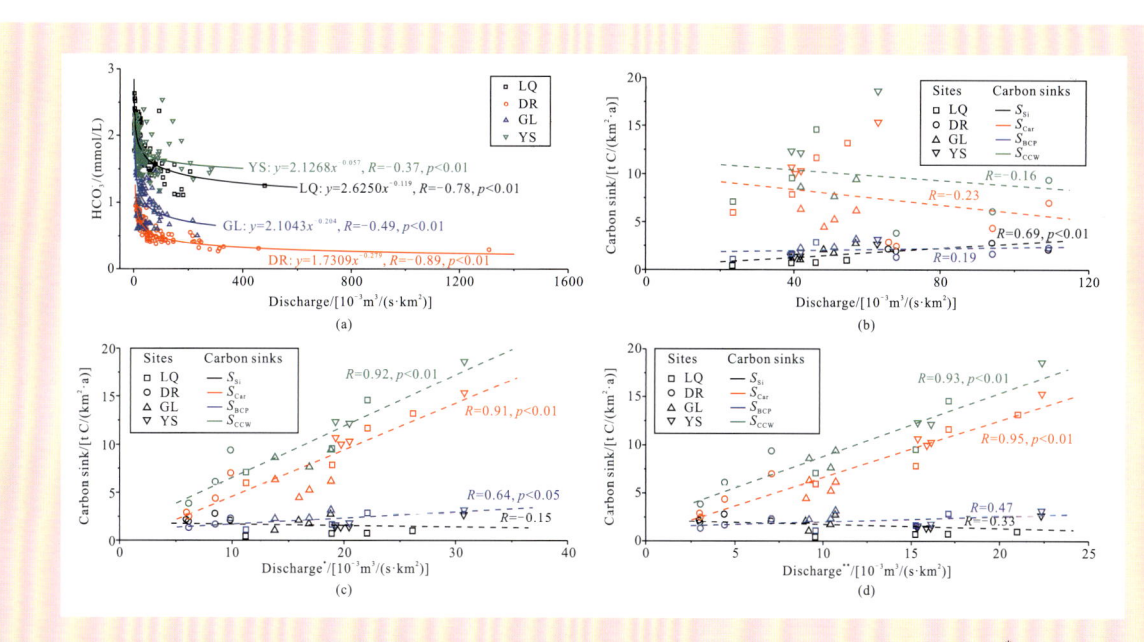

Plots of HCO_3^- vs. discharge (a), carbon sink vs. discharge (b), carbon sink vs. discharge* (c), and carbon sink vs. discharge** (d) in the four sites. (Discharge* = discharge × proportion of karst area; Discharge** = discharge × proportion of karst area × contribution of HCO_3^- sourced from carbonate weathering by carbonic acid)

四个观测点：（a）HCO_3^- 与流量；（b）碳汇与流量；（c）碳汇与流量*；（d）碳汇与流量**（流量* = 流量 × 流域岩溶区占比；流量** = 流量 × 流域岩溶区占比 × 碳酸风化碳酸盐岩对 HCO_3^- 的贡献比例）

5. Inorganic carbon cycle in reservoir fed by karst groundwater
岩溶地下水补给型水库无机碳循环

Most reservoirs in subtropical areas experience periodic variations in the thermal structure of their water columns, with times of strong thermal stratification being succeeded by periods of mixing, over the course of the year. Understanding of the transport and transformation of dissolved inorganic carbon over such thermal cycles in artificial reservoirs remains poor. To address this problem, this study examined the spatiotemporal behavior of dissolved inorganic carbon (DIC), the partial pressure of CO_2 (pCO_2), carbon isotope ratios ($\delta^{13}C_{DIC}$), and CO_2 emission (FCO_2), from 2014 to 2018 in a subtropical, groundwater-fed reservoir in southern China. It was found that CO_2 emissions during mixing periods are much higher than in thermally stratified periods (particularly during the transition from stratified to mixing) as a result of upwelling and release of dissolved CO_2 (CO_{2aq}) accumulated in the hypolimnion. CO_2 emission fluxes at the water-gas interface accounted for only a small proportion of the DIC in the reservoir. The relationships between DIC and $\delta^{13}C_{DIC}$ displayed two distinct modes, due to spatial differences in water depths and strong thermal stratification during warmer seasons: ① DIC concentrations increase and $\delta^{13}C_{DIC}$ values decrease from epilimnion to hypolimnion, and ② $\delta^{13}C_{DIC}$ values decrease with increasing DIC concentrations but $\delta^{13}C_{DIC}$ is progressively enriched near the bottom during periods of thermal stratification. In addition, this study found three distinct processes of DIC accumulation and consumption in the reservoir: ① DIC accumulated in the hypolimnion during thermal stratification periods, due to carbon retention but ② DIC was substantially consumed in the epilimnion during such periods, and ③ average DIC concentrations and pCO_2 increased significantly from upstream to downstream along the reservoir, while average $\delta^{13}C_{DIC}$ values became lighter. These results highlight that carbon behavior in groundwater-fed reservoirs is often controlled by a combination of biogeochemical processes and seasonal variations in thermal structure. Sampling and monitoring strategies (time and frequency) should consider these factors in order to accurately estimate carbon budgets and carbon sink effects in reservoirs, lakes or ponds (Li et al., 2022).

亚热带地区的大部分水库存在周期性的热结构变化，强烈的热分层之后是混合期。本研究聚焦岩溶水库热结构变化过程中的溶解无机碳的迁移与转化，以弥补对岩溶水库碳循环研究的不

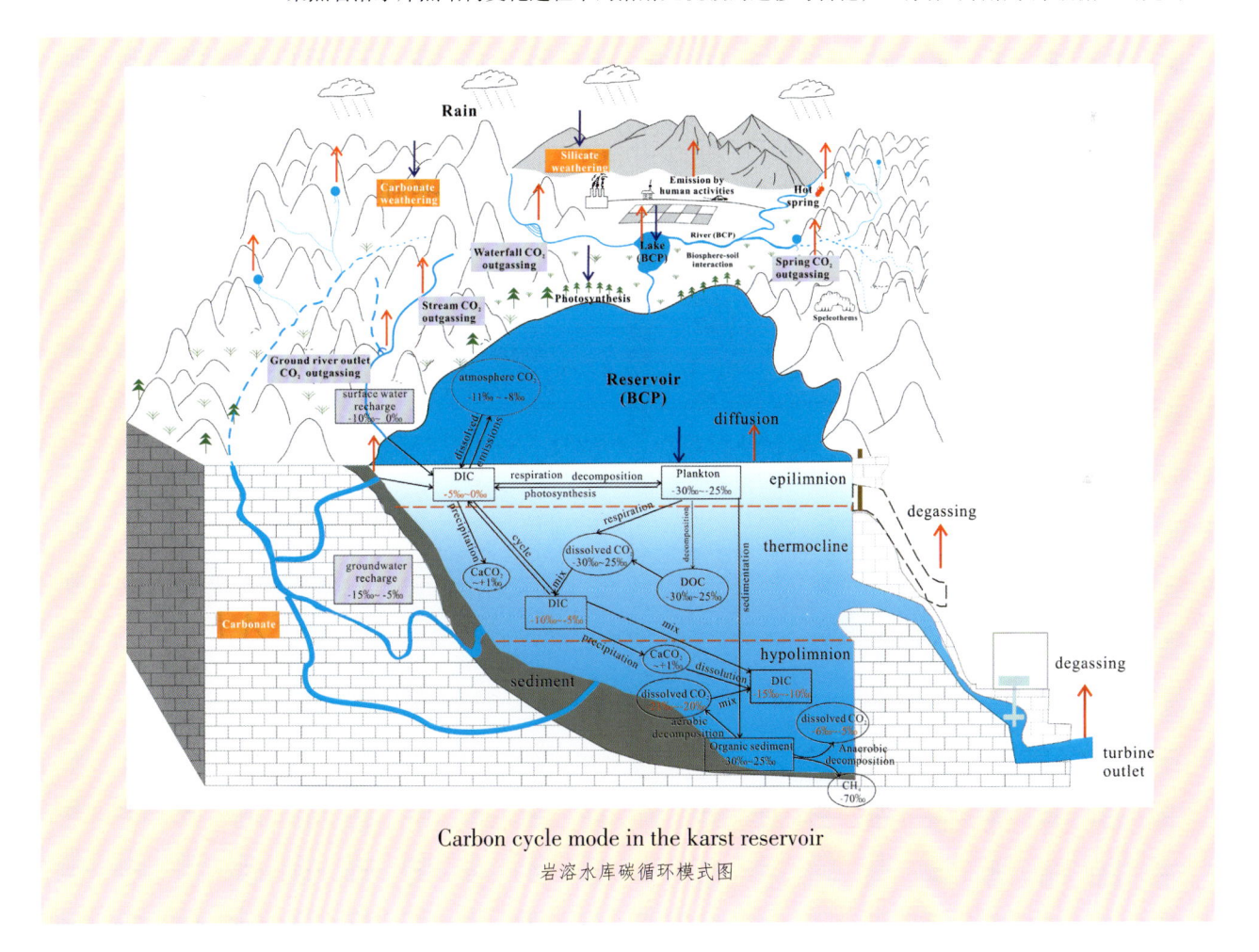

Carbon cycle mode in the karst reservoir
岩溶水库碳循环模式图

足。从 2014~2018 年，针对中国南方亚热带地区地下水补给的典型岩溶水库开展了溶解无机碳（DIC）、CO_2 分压（pCO_2）、稳定碳同位素比值（$\delta^{13}C_{DIC}$）和 CO_2 交换通量（FCO_2）的长期监测。发现 CO_2 在混合期的排放量要远高于热分层期，尤其是在分层期向混合期转化的过渡期，这是水库底水层积累的溶解二氧化碳 (CO_{2aq}) 上升和释放所致，但质量平衡研究表明，水—气界面 CO_2 排放通量仅占水库 DIC 总量的一小部分。由于水体深度空间变化及暖季强烈的热分层现象，溶解无机碳（DIC）和稳定碳同位素比值（$\delta^{13}C_{DIC}$）之间的关系展现出两种不同的模式：①从表水层到底水层，DIC 浓度增加，$\delta^{13}C_{DIC}$ 降低；②随着 DIC 浓度增加，$\delta^{13}C_{DIC}$ 降低，但在热分层期间，$\delta^{13}C_{DIC}$ 在接近底部区域逐步增加。此外，该研究还发现了水库中 DIC 的三种不同堆积与消耗方式：①由于碳阻滞，DIC 在热分层期间在底水层累积；②同期，DIC 在表层水中发生明显消耗；③ DIC 平均浓度和 CO_2 分压沿水库上游到下游明显增加，同时，$\delta^{13}C_{DIC}$ 均值变小。这些结果表明，地下水补给的水库中碳的行为通常受到生物地球化学过程和水库热结构季节变化的综合控制。因此，在采样和监测时应充分考虑这些因素，合理确定采样和监测时间及频率，以便准确评估水库的碳收支和碳汇效应（Li et al., 2022）。

6. Impact of nitrogen on karst carbon cycle in the Lijiang River Basin
漓江流域氮素对岩溶碳循环过程的影响机制

Through monitoring of the Lijiang River Basin, combined with a series of potted plant simulation experiments with different nitrogen application concentrations, the study has found that: ① Nitrogen fertilizer increased calcareous soil CO_2 by 10.5%–30.6%, and the dissolution rate of the tablets was increased by 1.8–3.6 times. The soil respiration rate also increased with the amount of fertilization, averaging 26.97 to 48.95 mg C/(m^2·h), which was increased by 7%–60% compared with that without nitrogen fertilization. Fertilization led to an increase in both carbon source and sink in the soil, and the sink/source ratio increased from 0.44% to 0.91% with the increase of the applied nitrogen fertilizer. ② There are three acid buffering mechanisms in the calcareous soils: carbonate dissolution by carbonic acid, carbonate dissolution by nitric acid, and cation exchange. The lower concentration of nitrogen fertilizer (100 kg N/(hm^2·a)) mainly participated in the karst carbon cycle by increasing the soil CO_2 concentration indirectly. The acid from nitrification was buffered by cation exchange, and the soil carbonate calcium dissolution mainly came from soil CO_2. When fertilization concentration is 250–700 kg N/(hm^2·a), 45% H^+ participated in the calcium carbonate dissolution directly, and 55% was buffered by cation exchange. The $\delta^{13}C_{DIC}$ of leachate was controlled by soil CO_2 partial pressure rather than the strength of nitrification. ③ The concentration of inorganic carbon, calcium, and magnesium in groundwater, derived from carbonate dissolution by carbonate acid, nitric acid, and cation exchange, increases with NO_3^- concentration in Lijiang River Basin. When the input of NO_3^- was less than 0.2 mmol/L, the process is dominated by the fully absorption of nitrogen by plants that stimulated microbial respiration and organic mineralization, and increased the soil CO_2 concentration to dissolve carbonate rocks; when the input of NO_3^- was more than 0.3 mmol/L, the process is dominated by the carbonate dissolution by nitric acid or cation exchange. Both the HCO_3^- concentration and $\delta^{13}C_{DIC}$ of underground rivers were controlled by CO_2 partial pressure. ④ In the underground river system, the average ratio of the carbonate dissolution by nitric acid calculated by the end member mixing model of isotope was 4.34%, while calculated by the water chemical equilibrium method was 8.83%. The difference of 4.49% between the two methods was probably entirely due to the cation exchange. The results may be helpful to improve the theory of carbon and nitrogen coupled cycle in karst dynamic systems, supporting data for accurate calculation of the influence on karst carbon cycle and karst carbon sink by nitrogen fertilizer application, and providing scientific support for the rational use of nitrogen fertilizer to reduce related pollution (Huang, 2020).

通过漓江流域的监测，结合不同施氮浓度的盆栽模拟实验研究发现，①氮肥对土壤 CO_2 的提高作用为 10.5%~30.6%，试片溶蚀速率提高了 1.8~3.6 倍。土壤呼吸速率也随施肥量增加而提高，平均值为 26.97~48.95 mg C/(m²·h)，比不施肥的提高了 7%~60%。施肥导致土壤碳源、汇量均增加，随施氮量的增加，汇/源比从 0.44% 上升到 0.91%。②石灰土存在碳酸溶解碳酸钙、硝酸溶解碳酸钙和阳离子交换三种酸缓冲机制。较低浓度的氮肥（100 kg N/(hm²·a)）主要通过增加土壤 CO_2 的浓度间接参与岩溶碳循环。硝化产酸全部由阳离子交换缓冲，土壤碳酸钙溶蚀全部来自土壤 CO_2。在施肥浓度为 250~700 kg N/(hm²·a) 时，45% 的 H^+ 直接参与碳酸钙的溶蚀，55% 的 H^+ 被阳离子交换缓冲。渗漏液 $\delta^{13}C_{DIC}$ 受控于土壤 CO_2 分压而不是硝化作用的强弱。③漓江流域地下水无机碳和钙、镁浓度随 NO_3^- 浓度的增加而增加，三者来源于碳酸溶蚀碳酸盐、硝酸溶蚀碳酸盐和阳离子交换三个过程。在人为输入的 NO_3^- < 0.2 mmol/L 时，以植物充分吸收氮素，刺激微生物呼吸和有机质矿化，增加土壤 CO_2 溶蚀碳酸盐为主；当 NO_3^- >0.3 mmol/L，以硝酸溶蚀或阳离子交换为主。地下河 HCO_3^- 浓度与 $\delta^{13}C_{DIC}$ 均受 CO_2 分压控制。④同位素端元法计算的地下河硝酸溶蚀碳酸盐平均值为 4.34%，水化学平衡法计算的结果为 8.83%，这 4.49% 的差值可能全部为阳离子交换造成。研究结果有助于完善岩溶动力系统碳氮耦合循环理论，为准确计算氮肥施用对岩溶碳循环和岩溶碳汇的影响提供数据支撑，同时为合理利用氮肥减少氮污染提供科学支持（黄芬，2020）。

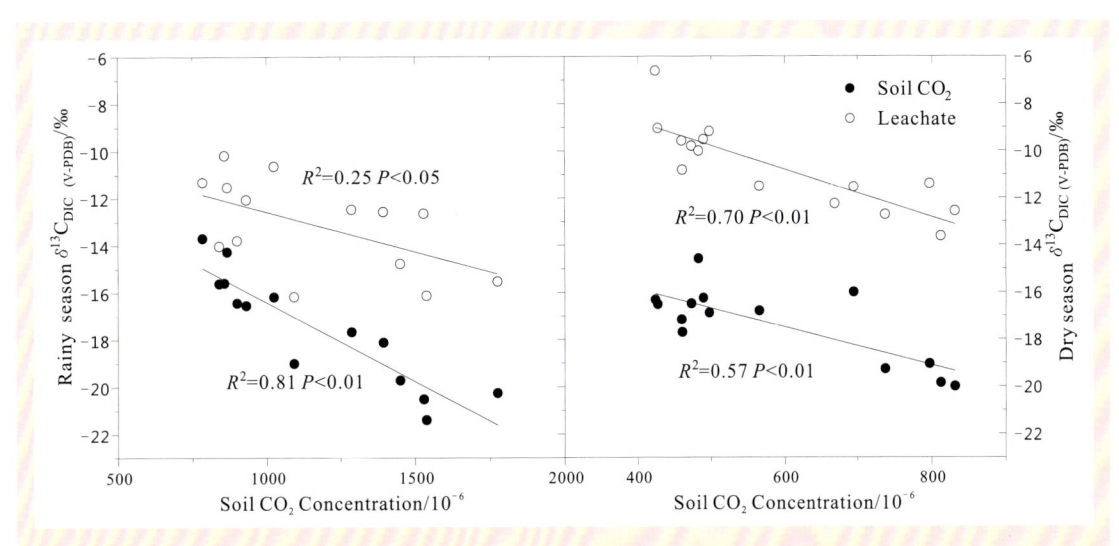

Soil CO₂ concentration and $\delta^{13}C_{DIC}$ - CO₂ and leachate $\delta^{13}C_{DIC}$ relationship

土壤 CO₂ 浓度与 $\delta^{13}C$–CO₂ 和渗滤液 $\delta^{13}C_{DIC}$ 的关系

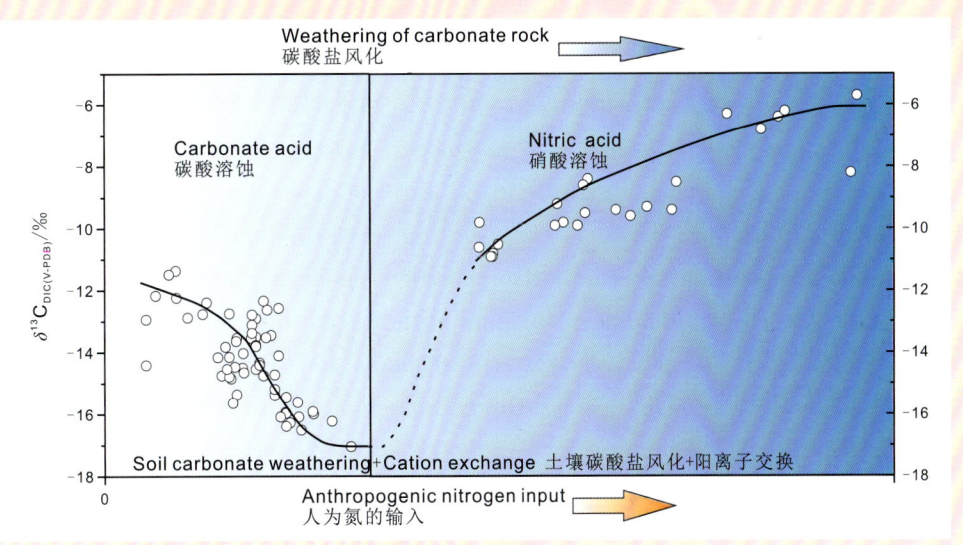

Sketch map of the influence on $\delta^{13}C_{DIC}$ by anthropogenic nitrogen input in the river basin

人为氮输入对流域 $\delta^{13}C_{DIC}$ 的影响示意图

3.2.1.2 Karst sediments for paleoclimate reconstruction
岩溶沉积物重建古气候

1. Climatic controls on travertine deposition in southern Xizang during the late Quaternary

气候对西藏南部第四纪钙华沉积的控制作用

Large volumes of travertine deposits are preserved at hydrothermal spring sites on the Tibetan Plateau (TP). Yet, most of these deposits are under-researched with respect to their diagenetic and depositional history, and there is still very limited understanding of the tectonic and climatic influences on travertine precipitation in the arid high-altitude setting of Xizang. In this study, a detailed uranium-series dating campaign was carried out for the Qiusang travertine (~4,270 m above sea level), southern Xizang that has been previously dated back to 486 thousand years ago (ka). Based on 42 new ^{230}Th/U ages, combined with geomorphological and sedimentological investigations, seven travertine zones were identified, and distinct travertine depositional phases were constrained. 11.7-6.8 ka and ~13.4 ka (zone 1), 128-122 ka (zone 2), ~193 ka (zone 3), ~292 ka and 324 ka (zone 4), >317 ka (zone 5), ca. 415 to 470 ka (zone 6), and ca. 419 to 445 ka (zone 7). Comparison of these depositional phases with local and regional proxy records suggests that travertine accumulation at Qiusang occurred during main interglacials when monsoon precipitation peaked on the TP. This coincidence, together with a sensitive response of Tibetan hydrothermal spring activity to meteoric recharge, implies that climate controls the precipitation of large travertine volumes on orbital timescales on the plateau. The study proposes that (i) tectonic activity is of subordinate importance and influences travertine precipitation on the TP only episodically and on significantly shorter (i.e. centennial to millennial) timescales related to the recurrence rates of large earthquakes and that (ii) intensive monsoonal-driven groundwater recharge is required on top of tectonic activity for generating volumetrically significant travertine accumulations. Because of the high precipitation rates typical for hydrothermal spring carbonates, the study concludes that travertine deposits on the TP could be utilized as valuable high-resolution proxy records of peak monsoon conditions in the currently arid to semi-arid landscape. Furthermore, the Qiusang travertine zone 7 is terraced and the travertine layers adjusted to a paleo-riverbed elevation ~30 m above the current river, allowing us to constrain fluvial incision to ~0.07 m/ka for the south-central sector of the TP since the Mid-Pleistocene. The abundant travertine occurrences in Tibet in combination with uranium-series dating can thus also provide detailed insights into earth surface dynamics and landscape evolution on the world's highest plateau (Wang et al., 2022).

青藏高原上分布着许多由地热泉水形成的钙华。目前，关于这些钙华沉积的成因及发育历史的研究比较少，对构造和气候在这一干旱、高海拔地区钙华沉积演化中的作用的认识还不深入。本研究对西藏南部最大的钙华——邱桑钙华（海拔约 4270 m）进行了系统的铀系测年研究，以往测年表明该钙华的年代可追溯至 48.6 万年前。基于 42 个新 ^{230}Th/U 年龄，结合详细的地貌和沉积学调查，共确定了 7 个钙华沉积子区，每个子区形成于不同的时期：1.17 万 ~0.68 万年前和约 1.34 万年前（子区 1）、12.8 万 ~12.2 万年前（子区 2）、约 19.3 万年前（子区 3）、约 29.2 万年和 32.4 万年前（子区 4）、>31.7 万年前（子区 5）、约 41.5 万 ~47 万年前（子区 6）、约 41.9 万 ~44.5 万年前（子区 7）。将这些沉积时段与当地和区域的古气候记录对比发现，邱桑钙华沉积主要发生于间冰期阶段，此时增强的印度夏季风使得高原降水显著增加。结合西藏地热泉活动对大气降水补给的敏感性分析，表明轨道时间尺度上的气候变化可能在很大程度上控制了大规模钙华的发育和演化。而构造活动对钙华发育的影响具有次要性，地震活动也仅能在相对较短的时间尺度（即百年至千年）上影响热泉活动和钙华沉积。在构造影响的基础上，需要有季风增强带来的大量大气降水补给，才会产生强烈的地下热泉活动，继而在地表沉积大规模的钙华。鉴于热成因钙华较高的沉积速率，青藏高原上的热成因钙华沉积的发生有很大潜力成为当前干旱至半干旱地区且受季风影响显著的青藏高原上的古气候代用指标之一。此外，邱桑钙华沉积子区 7 发育在古河流阶地上，通过古河床高程与现代河面的高差（约 30 m）和阶地上钙华沉积时代可以估算出中更新世以来青藏高原中南部的河流下切速率大约为 0.07 m/ka。这也表明广泛分布的钙华沉积凭借其铀系测年优势在世界最高高原地表过程和地貌演变研究中也具有重要应用价值（Wang et al., 2022）。

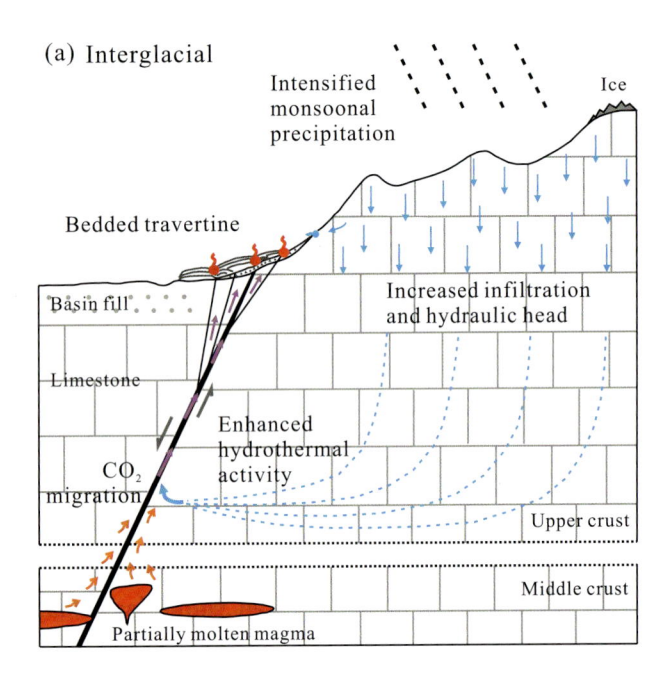

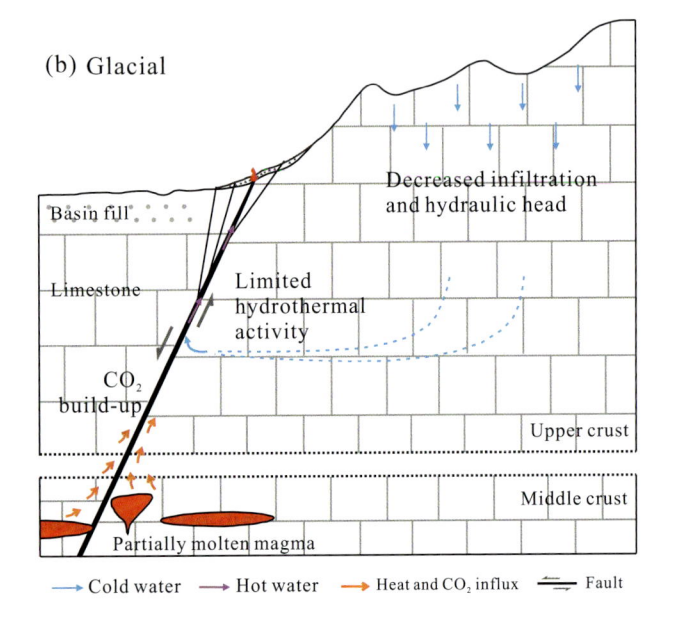

Conceptual model (partly based on Uysal et al., 2019) illustrating the climate-controlled hydrothermal activity and travertine deposition in southern Xizang

概念模型（部分基于 Uysal et al., 2019）说明了西藏南部气候控制的热液活动和钙华沉积

2. Deciphering the hydroclimatic significance of dripwater $\delta^{13}C_{DIC}$ variations in monsoonal China based on modern cave monitoring

洞穴滴水 $\delta^{13}C_{DIC}$ 变化揭示中国季风区水文气候意义

Stable carbon isotopic composition ($\delta^{13}C$) of speleothems has often been recognized as a proxy of vegetation and soil processes in many climatic regimes. In monsoonal regions, the speleothem $\delta^{13}C$ records are thought to be able to document changes in local and/or regional precipitation as well. Due to the complexity of carbon isotopic evolution within heterogeneous karst aquifers, accurate interpretations of this proxy are challenging. It is essential to carry out detailed monitoring of cave dripwater $\delta^{13}C_{DIC}$ to disentangle various processes governing carbon isotopic evolution and decipher the hydroclimatic constraints. Here, the study reported the results of a five-year monitoring of the $\delta^{13}C_{DIC}$ as well as hydrochemical compositions of dripwaters in the Maomaotou Big Cave, Guilin, South China. Great spatiotemporal variations in the dripwater $\delta^{13}C_{DIC}$ values were observed: ① seepage flow-fed drips commonly had lower $\delta^{13}C_{DIC}$ values, with smaller temporal variation; ② the lowest $\delta^{13}C_{DIC}$ values were found at fracture-fed drips that had a large supply of soil CO_2, whereas the fracture-fed drips with less, episodic CO_2 recharge had much more positive $\delta^{13}C_{DIC}$; ③ the $\delta^{13}C_{DIC}$ exhibits remarkable season changes, particularly at fracture-fed drips where higher $\delta^{13}C_{DIC}$ values were observed in winter seasons; and ④ an increasing trend in mean value of $\delta^{13}C_{DIC}$ at individual sites during 2015–2019 was found. Both hydrochemical and isotopic analyses revealed that the dripwater $\delta^{13}C_{DIC}$ is mainly controlled by spatiotemporal variations in water-CO_2-rock interactions in association with hydrological processes, coupled with soil CO_2 dynamics, which are closely linked to changes in local rainfall at monthly to annual timescales. Prior calcite precipitation (PCP) along the flow path above the cave could also affect the $\delta^{13}C_{DIC}$ of fracture-fed dripwaters in dry periods. Comparisons of the dripwater $\delta^{13}C_{DIC}$ records with local climate showed that the change of summer monsoonal rainfall, particularly the August-September-October precipitation, significantly influences the annual mean $\delta^{13}C_{DIC}$ value of dripwaters in caves of South China. Further comparison analyses of dripwater $\delta^{13}C_{DIC}$ records from both southern and northern China suggested that variations of dripwater $\delta^{13}C_{DIC}$ could reflect changes of regional monsoonal precipitation over inter-annual (and maybe decadal) timescale. This demonstrates that the $\delta^{13}C$ of speleothems, if precipitated at isotopic equilibrium and not greatly influenced by the CO_2 degassing, is likely to be a valuable proxy of monsoon rainfall variability across East Asia (Yin et al., 2021).

在许多气候条件下，洞穴次生化学沉积物的稳定碳同位素组成（$\delta^{13}C$）通常被认为是植被和土壤演化过程的环境指标。在季风区，洞穴次生化学沉积物的 $\delta^{13}C$ 记录被认为能够记录当地和/或区域降水量的变化。由于非均质岩溶含水层中碳同位素演化的复杂性，对这一指标的准确解释具有挑战性。因此，有必要对洞穴滴水 $\delta^{13}C_{DIC}$ 进行详细监测，以理清控制碳同位素演化的各种过程，并解译水文气候制约因素。本研究获取了中国南方桂林茅茅头大岩 5 年洞穴滴水的 $\delta^{13}C_{DIC}$ 值及水化学组成。监测结果表明，茅茅头大岩洞穴滴水的 $\delta^{13}C_{DIC}$ 值时空变化巨大：①渗流补给的滴水通常具有较低的 $\delta^{13}C_{DIC}$ 值，其时间变化较小；②裂隙水补给滴水，当含有较大量的由土壤供给的 CO_2 时，其 $\delta^{13}C_{DIC}$ 值最低，当 CO_2 的供应量偏少且不稳定时，其 $\delta^{13}C_{DIC}$ 偏正；③ $\delta^{13}C_{DIC}$ 表现出明显的季节性变化趋势，尤其是裂隙水补给滴水，具有较高的 $\delta^{13}C_{DIC}$ 值；④ 2015~2019 年，各观测点的 $\delta^{13}C_{DIC}$ 平均值呈上升趋势。水化学和同位素分析均表明，滴水的 $\delta^{13}C_{DIC}$ 主要受与水文过程相关的水 –CO_2– 岩石相互作用的时空变化及土壤 CO_2 动态的控制，而土壤 CO_2 动态与月至年的当地降水量变化密切相关。在旱季，沿着洞穴上方流动路径的已析出的方解石也会影响裂隙水补给滴水的 $\delta^{13}C_{DIC}$。洞穴滴水 $\delta^{13}C_{DIC}$ 记录与当地气候的比较表明，夏季季风降水的变化，尤其是 8、9、10 月降水，对中国南方洞穴滴水的 $\delta^{13}C_{DIC}$ 年平均值有显著影响。对中国南方和北方洞穴滴水的 $\delta^{13}C_{DIC}$ 记录的进一步对比分析表明，洞穴滴水的 $\delta^{13}C_{DIC}$ 变化可以反映区域季风降水在年际（可能是十年）尺度上的变化。这表明，如果在同位素平衡条件下沉淀，并且不受 CO_2 脱气的影响，洞穴滴水的 $\delta^{13}C$ 可能是整个东亚季风降雨变化的一个极具价值的环境指标（Yin et al., 2021）。

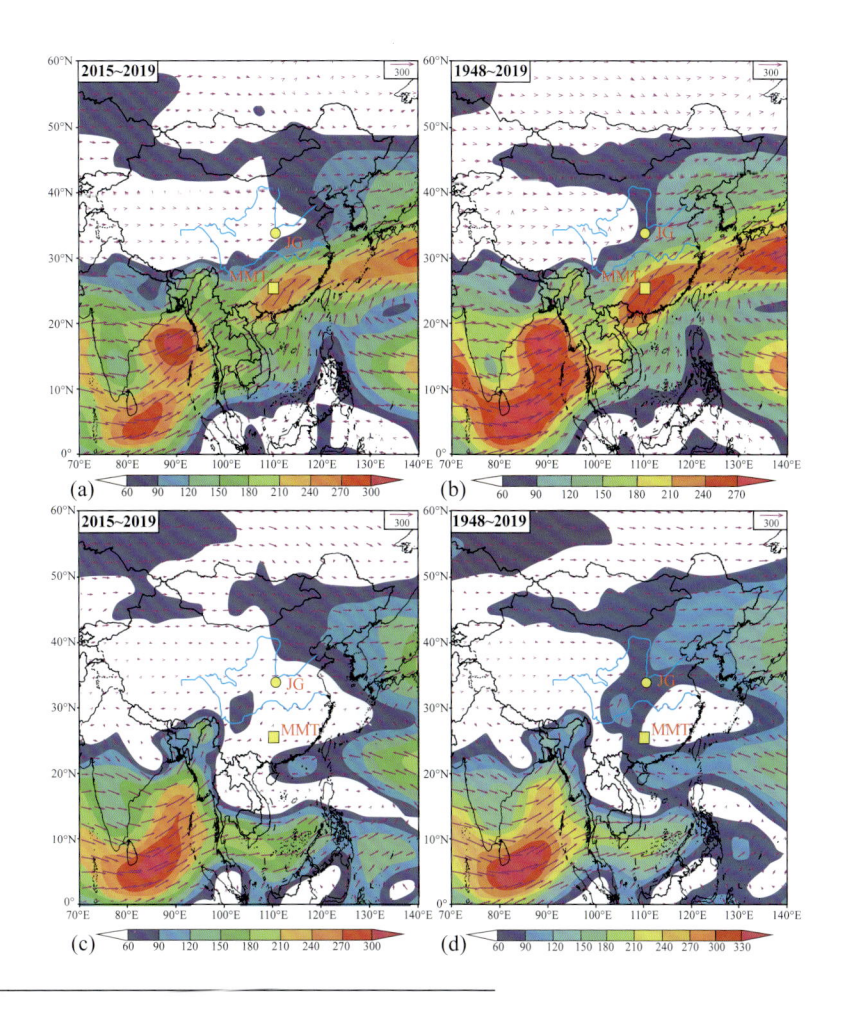

Integrated mean vapor flux (from surface to 300 mbar) in April–May–June–July (AMJJ; images (a) and (b)) and August–September–October (ASO; images (c) and (d)). The cycles and squares indicate the location of Jiguan Cave (JG) and Maomaotou Big Cave (MMT), respectively. It can be noted that the mean AMJJ vapor flux of East China in 2015–2019 is similar to that of 1948–2019, whereas there is a distinct difference in the mean ASO vapor flux between periods of the 2015–2019 and 1948–2019. More ASO vapor transport to northern China driven by the summer monsoon is usually associated with more rainfall in northern China.

4、5、6、7月（AMJJ；图像(a)和(b)）和8、9、10月（ASO；图像(c)和(d)）的综合平均水汽通量（从地表大气压到300mbar）。圆圈和方框分别表示鸡冠洞（JG）和茅茅头大岩（MMT）的位置。2015~2019年中国东部的4~7月平均水汽通量与1948~2019年相似，而2015~2019年和1948~2019年的8~10月平均水汽通量存在明显差异。8~10月，夏季风驱动更多水汽向中国北方输送，通常与中国北方降雨量更多相关。

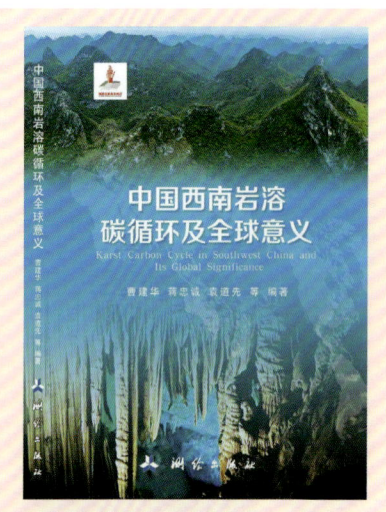

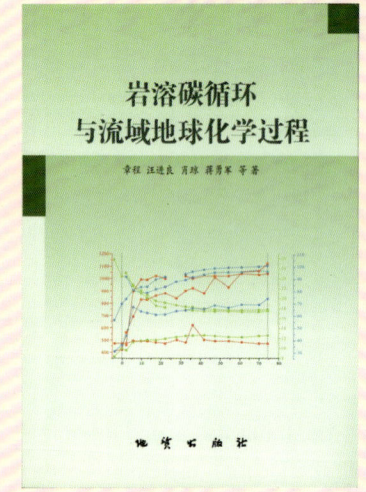

Published 4 monographs: *Modern Karstology*, *Karst Carbon Cycle in Southwest China and Its Global Significance*, *Karst Carbon Cycle and Geochemical Process in Catchments*, and *Karst Cave Environment and Stalagmite Paleoclimate Records*

出版专著4部：《现代岩溶学》《中国西南岩溶碳循环及全球意义》《岩溶碳循环与流域地球化学过程》《岩溶洞穴环境及石笋古气候记录》

第三章　科学研究

The popular science article on karst carbon sink published on Xuexi Qiangguo platform
岩溶碳汇成果进入"学习强国"平台

Geology and Mineral Resources industrial standard of China: *Guidance for karst carbon cycle survey and carbon sink evaluation*
编制的中国地质矿产行业标准《岩溶碳循环调查与碳汇效应评价指南》正式施行

3.2.2 Karst water resources and water security
岩溶水资源与水安全

1. Karst water environment in Southeast Asia
东南亚岩溶水环境

The carbonate areas of Southeast Asia are part of the global set of well-developed tropical-subtropical karst regions and form water-rich aquifers. Due to the strong development of karst features, groundwater in karst conduits flows rapidly and is susceptible to various environmental problems, including rocky desertification and socioeconomic impacts leading to poverty. Karst-related data for the region are scarce and scattered. Based on information contributed by training workshops of the International Research Centre on Karst (IRCK) under the auspices of UNESCO, as well as published literature, this study summarizes karst hydrogeological data and water-related environmental issues in Southeast Asia, in an attempt to find commonality, and to form both generally valid and region-specific concepts that can be extended to data-deficient areas, where these concepts may serve as a guide for governments when managing the karst environment. Based on topographic differences, karst terrains in Southeast Asia were classified into four types: karst on plateaus, karst in mountains, karst in plains, and karst on islands. The examples shared by participants in the IRCK training workshops included karst information from their own countries, most of which have not been published in English. The case studies demonstrate that karst areas in Southeast Asia are widely and repeatedly exposed to droughts and floods, resulting in environmental constraints and development obstacles. These studies also show that environmental problems can be resolved and sustainable development can be achieved if appropriate management measures are taken (Jiang et al., 2021).

从全球角度来看，东南亚碳酸盐岩分布地区不仅具有发育良好的热带、亚热带岩溶形态，而且形成了富水的含水层。由于岩溶发育强度高，岩溶管道中的地下水流动迅速，容易受到各种环境问题的影响，并且导致石漠化和贫困等社会经济问题。东南亚地区的岩溶相关数据较少且不成体系。该研究基于联合国教科文组织国际岩溶研究中心培训班学员提供的各国岩溶信息和相关文献，将东南亚岩溶水文地质数据和水环境数据进行汇总和对比分析，找出共性，形成既有普遍适用性，又具有地区特点的科学概念，可推广至缺少数据的区域，从而为当地政府管理岩溶环境提供指导。根据地形差异，东南亚岩溶地形被分为4类：高原岩溶、山地岩溶、平原岩溶和岛屿岩溶。中心培训班学员分享的研究案例来自本区各个国家，且多未以英文形式发表。案例研究表明，东南亚岩溶区广泛且频繁遭受旱涝灾害，形成环境制约和发展障碍。这些研究同样表明，如采取恰当的管理策略，相关环境问题能得以解决，实现可持续发展（Jiang et al., 2021）。

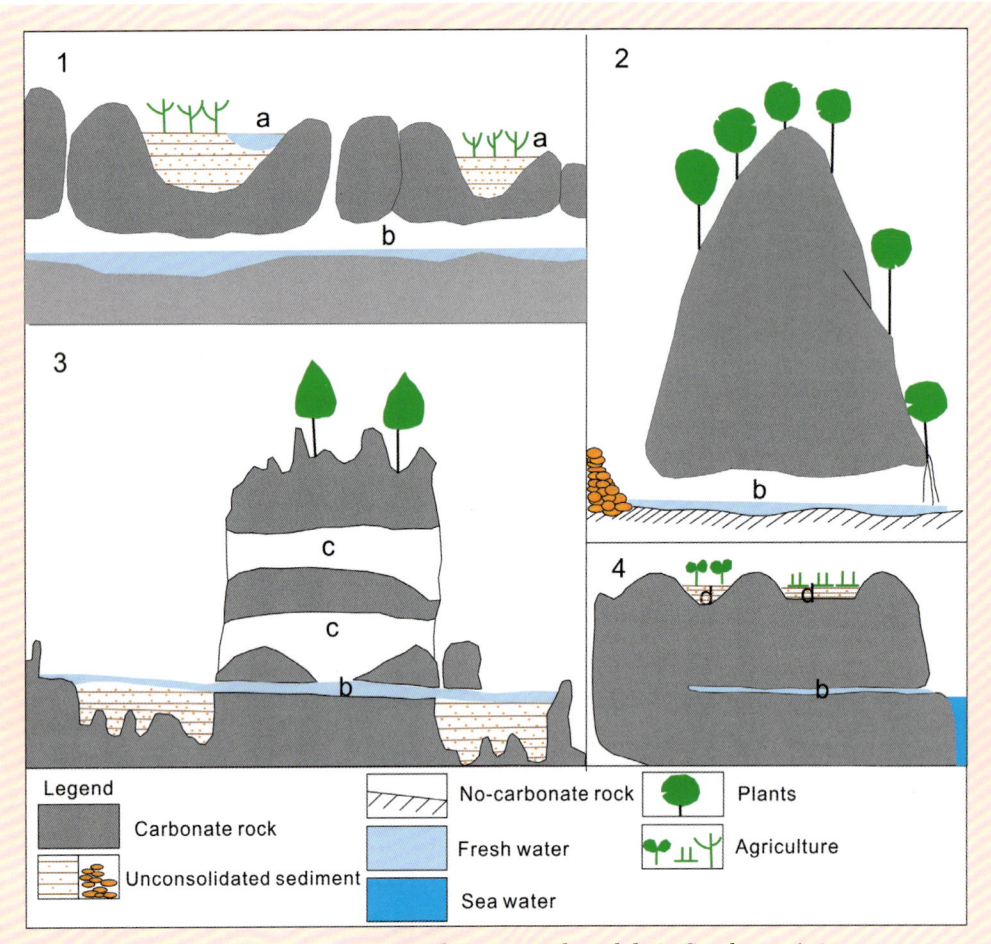

Four karst water environmental conceptual models in Southeast Asia
东南亚的四种岩溶水环境概念模型

1. Plateau polje model, based on the Yunnan plateau (Yuan and Cai, 1988); 2. Mountain model, based on Guangxi (Yuan, 1997); 3. Plain model, based on the Niah National Park (Dodge-Wan et al., 2017); 4. Island model, based on Gunung Sewu (Urushibara-Yoshino, 1991). a. Polje; b. Subterranean river; c. Cave; d. Doline

1. 高原坡立谷模型，基于云南高原（袁道先和蔡桂鸿，1988）；2. 山地模型，基于广西（Yuan, 1997）；3. 平原模型，基于尼亚国家公园（Dodge-Wan et al., 2017）；4. 岛屿模型，基于色乌山（Urushibara-Yoshino, 1991）；a. 坡立谷；b. 暗河；c. 洞穴；d. 溶蚀漏斗

2. Hydrochemical features and their controlling factors in the Kucaj-Beljanica Massif, Serbia
塞尔维亚 Kucaj-Beljanica 地块水化学特征及其控制因素

The Kucaj-Beljanica massif represents a complex hydrological karst system that possesses enormous potential for further groundwater extraction and regional water supply uses. In this paper, the influencing factors on groundwater quality in this area were discussed using hydrochemical analysis and the factor analysis method with 123 water and 20 rain/snow water samples. Hydrochemical analysis of the cation composition of cold springs, thermal springs/wells, and sinkhole waters in the test area indicates that Ca^{2+} and Mg^{2+} are the dominant cations, representing 73%–98% of ion equivalent. The anion composition of water indicates that HCO_3^- is the dominant anion, represented by 73%–91% of ion equivalent. Only in the No.30 thermal spring is characterized by $K^+ + Na^+$ and $HCO_3^- + SO_4^{2-}$. The P_{CO_2} of thermal springs/wells, cold springs and sinkhole waters decreases gradually by an average of 10,247 ppmv, 3,444 ppmv, and 319 ppmv (part per million of volume ratio), respectively. The average $\delta^{13}C_{DIC}$ of thermal springs/wells, cold springs, and sinkhole waters is −6.56‰, −10.19‰, and −13.46‰, respectively. Corresponding to $\delta^{13}C_{DIC}$, the mole ratios of $(Ca^{2+} + Mg^{2+})/HCO_3^-$ are 0.48, 0.55, and 0.60, respectively. Factor analysis identifies 3 sources of solutes: ① precipitation; ② water-rock interactions; and ③ soil leaching. Ions of Na^+, K^+, Cl^-, and SO_4^{2-} indicate the predominant influence of atmospheric precipitation, but the No.30 thermal spring sample is probably the result of deep water-rock interaction (volcanic rock) or is influenced by connate water mixed with shallow karst water. The dissolution of carbonate rocks is the primary factor affecting the Ca^{2+}, Mg^{2+}, and HCO_3^- contents of groundwater, and soil leaching is the primary factor controlling the concentration of NO_3^- in water. These results provide a scientific basis for rational exploitation, protection and land use planning in the test area (Huang et al., 2019).

Kucaj-Beljanica 地块是一个复杂的岩溶水文系统，具有进一步开采地下水和区域供水的巨大潜力。采用水化学分析和因子分析法，以 123 份水和 20 份雨雪水样为研究对象，探讨了影响该区地下水水质的因素。试验区冷泉、热泉/井、落水洞的水化学分析表明，Ca^{2+}、Mg^{2+} 为优势阳离子，占离子当量的 73%~98%。HCO_3^- 为优势阴离子，占离子当量的 73%~91%。在 30 号温泉中，其特征为 K^++Na^+ 和 $HCO_3^-+SO_4^{2-}$。热泉/井、冷泉和落水洞水的二氧化碳平均分压分别为 10247 ppmv、3444 ppmv、319 ppmv（百万分体积比）。三者 $\delta^{13}C_{DIC}$ 平均值分别为 −6.56‰、−10.19‰ 和 −13.46‰，对应于 $\delta^{13}C_{DIC}$，三者 $(Ca^{2+}+Mg^{2+})/HCO_3^-$ 摩尔比值分别为 0.48、0.55 和 0.60。因子分析确定了溶质的 3 个来源：①降水；②水－岩相互作用；③土壤淋滤。Na^+、K^+、Cl^-、SO_4^{2-} 离子表明，大气降水是主要的溶质来源，而 30 号温泉样品可能是深部水－岩相互作用（火山岩）的结果，也可能是原生水与浅层岩溶水混合作用的结果。碳酸盐岩溶蚀作用是影响地下水中钙、镁、HCO_3^- 含量的主要因素，而土壤淋滤作用是控制水中 NO_3^- 浓度的主要因素。研究结果为试验区合理开发、保护和土地利用规划提供了科学依据（Huang et al., 2019）。

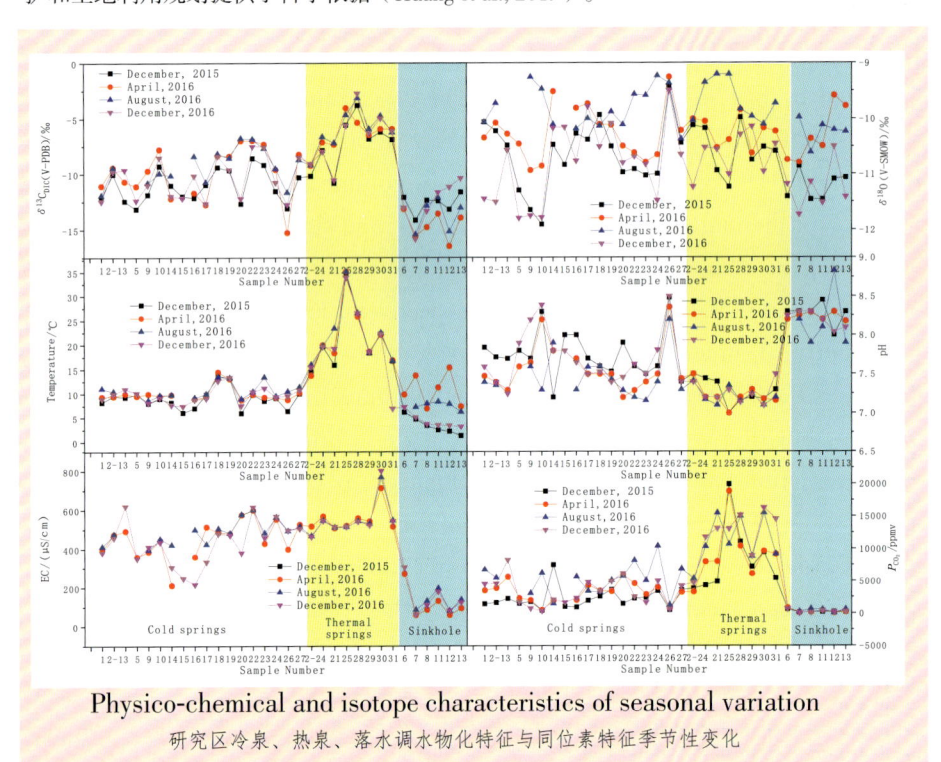

Physico-chemical and isotope characteristics of seasonal variation
研究区冷泉、热泉、落水调水物化特征与同位素特征季节性变化

3. Physicochemical parameters and phytoplankton as indicators of the aquatic environment in karstic springs of South China
中国南方岩溶泉作为水环境指示的物理化学参数和浮游植物指标

As the concentrated discharge outlet of an aquifer or groundwater system, a karst spring is partly independent from the aquifer, due to its formation of a pool or lake after outcropping onto the surface. Understanding how to evaluate the unique and sensitive environment of the karst spring is essential for water resource protection. Five karst springs in South China were investigated by analyzing their hydrodynamic conditions, variations in physicochemical parameters, and phytoplankton community structures. Dominated by regional or local ground-water flow, these springs had different catchment area characteristics and hydrogeological conditions. The results showed that, although they had similar water quality, their physicochemical parameters needed to be distinguished and evaluated in different ways in order to determine the cause of the observed degradation in spring water quality. Ca^{2+}, HCO_3^-, and specific electrical conductivity were the major parameters reflecting the impact of regional flow from the aquifer; pH, dissolved oxygen, and water temperature indicated the local environment in and around the springs; while nitrogen and COD_{Mn} both related to the aquifer and local environment, depending on seasonal variation and human activities. The comparison of long-term nitrate data revealed that environmental pressure has increased over time. The water deterioration of Lingshui Spring was attributed to the strong interaction of surface water and groundwater. High nutrient concentrations did not correspond with the highest phytoplankton abundance or the most species. The phytoplankton community structures in the karst springs varied from place to place, depending on the hydrogeological conditions and the surrounding environment. The water environment status, as reflected by the combination of water quality indices and biological indicators, could more comprehensively represent overall water health (Guo et al., 2019).

作为含水层或地下水系统的集中排泄口，岩溶泉在地表出露后形成池塘或湖泊，因此在一定程度上独立于含水层。了解如何评价岩溶泉独特而敏感的环境，对水资源保护至关重要。通过水动力条件、物理化学参数变化和浮游植物群落结构的分析，对南方五个岩溶泉进行了研究。受区域或局部地下水流量的控制，这些泉水具有不同的流域面积和水文地质条件。结果表明，尽管它们的水质相似，但它们的物理化学参数需要以不同的方式进行区分和评估，以确定泉水水质退化的原因。Ca^{2+}、HCO_3^-和电导率是反映含水层区域地下水流动的主要参数；pH、溶解氧和水温显示泉口周围的局部环境；而氮和COD_{Mn}均与含水层和局部环境有关，取决于季节变化和人类活动。对长期硝酸盐观测数据的比较表明，环境压力随着时间的推移而增大。灵水岩溶泉的水质恶化是地表水与地下水强烈相互作用的结果。高营养物浓度与最高的浮游植物丰度或最多的物种并不对应。各岩溶泉的浮游植物群落结构不同，与水文地质条件和周围环境有关。水质指标和生物指标相结合所反映的水环境状况可以更全面地反映岩溶泉整体的健康状况（Guo et al., 2019）。

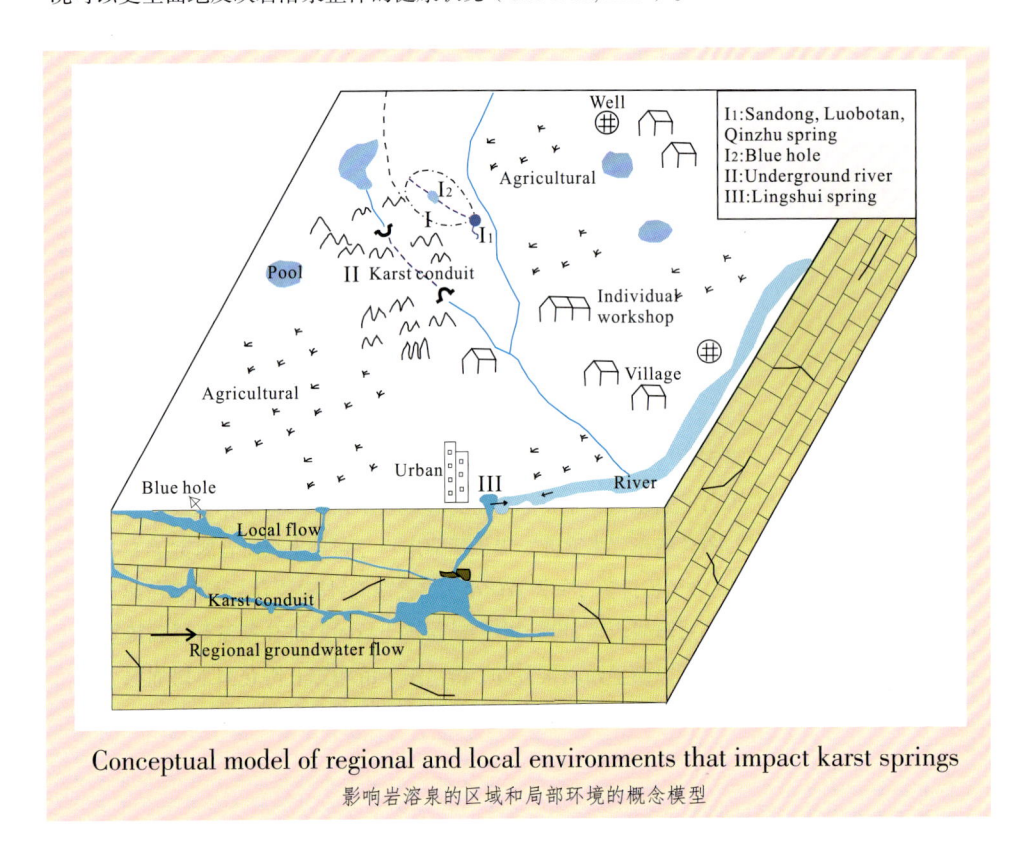

Conceptual model of regional and local environments that impact karst springs
影响岩溶泉的区域和局部环境的概念模型

4. Multi-geophysical approaches to detect karst channels underground –A case study in Mengzi of Yunnan Province, China
地球物理综合探测地下岩溶管道——以云南蒙自为例

Mengzi is located in the south 20 km away from the outlet of Nandong subsurface river and has been suffering from water deficiency in recent years. It is necessary to find out the water resources underground according to the geological characteristics such as the positions and buried depths of the underground river to improve the civil and industrial environments. Due to the adverse factors such as topographic relief, and bare rocks in karst terrains, the geophysical approaches, such as Controlled Source Audio Magnetotellurics and Seismic Refraction Tomography, were used to roughly identify faults and fracture zones by the geophysical features of low resistivity and low velocity, and then used the mise-a-la-masse method to judge which faults and fracture zones should be the potential channels of the subsurface river. Five anomalies were recognized along the profile of 2.4 km long and showed that the northeast river system has several branches. Drilling data have proved that the first borehole indicated a water-bearing channel by a characteristics of rock core of river sands and gravels deposition, the second one encountered water-filled fracture zone with abundant water, and the third one exposed mud-filled fracture zone without sustainable water. The results from this case study show that the combination of Controlled Source Audio Magnetotellurics, Seismic Refraction Tomography, and mise-a-la-masse is one of the effective methods to detect water-filled channels or fracture zones in karst terrains (Gan et al., 2017).

蒙自地处南洞地下河出口南部 20 km 处，近年来水资源短缺严重。有必要根据地下河位置和埋深等地质特征，获取岩溶地下水资源，以保障生活和工业用水安全。由于受地形起伏、岩溶区岩石裸露等不利因素影响，采用了可控源音频大地电磁法和地震折射层析成像法，通过低阻、低速的地球物理特征，大致识别断层和断裂带，然后利用充电法判断哪些断层和断裂破碎带异常由潜在的地下河管道所引起。在长 2.4 km 的剖面上发现了五个异常，表明在北东方向上地下河管道存在几条分支。钻探揭露：第一个钻孔钻遇地下河，以河砂、砾石沉积为特征，显示了一条富水管道；第二个钻孔揭示了富水的充水断裂带；第三个钻孔显示弱含水的充泥断裂带。本案例研究表明，可控源音频大地电磁法、地震折射层析法和充电法相结合是探测岩溶地区充水管道或断裂带最为有效的组合方法（Gan et al., 2017）。

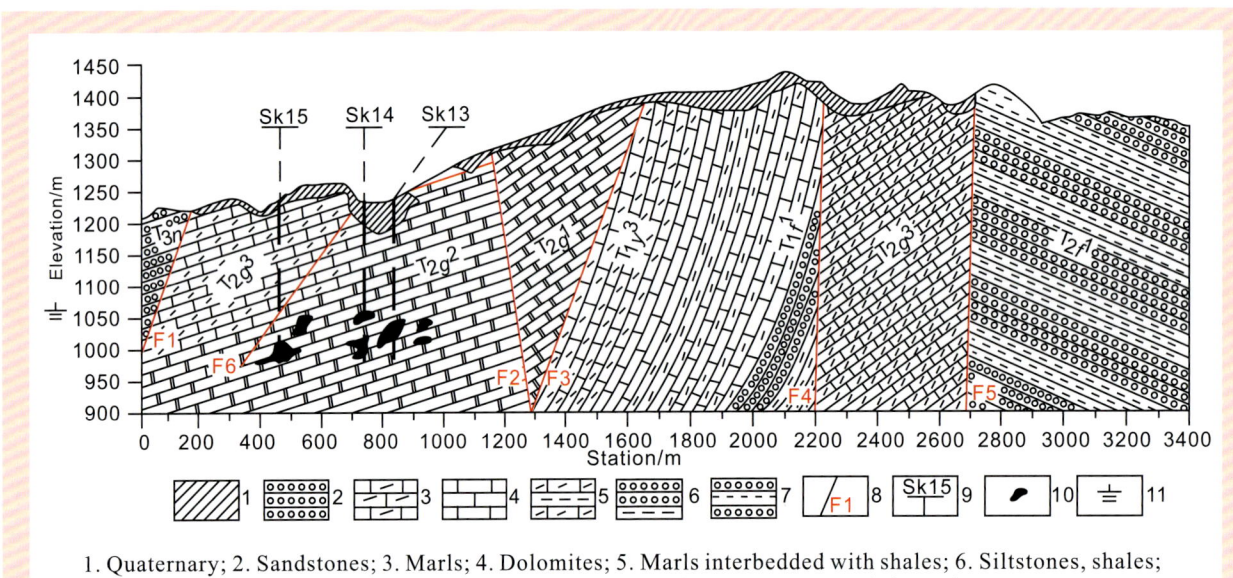

1. Quaternary; 2. Sandstones; 3. Marls; 4. Dolomites; 5. Marls interbedded with shales; 6. Siltstones, shales; 7. Shales, siltstones; 8. Fault; 9. Borehole and reference number; 10. Interpreted channel; 11. Water table at discharging outlet

Geological interpretation section
地质解译剖面

Integrated geophysics, drill hole locations and hydrogeological data
综合物探、钻孔及水文地质数据

Station no. 点位 /m			480	740	850
Anomalies in geophysics 地球物理异常	CSAMT (resistivities) 可控源音频大地电磁测深法（电阻率）		Lower 低	Lower 低	Lower 低
	Seismic refraction tomography(velocities) 地震折射速度成像法		Lower 低		
	Mise-a-la-masse (before filtering) 充电法（滤波前）	Potentials 电位	Peaks 峰值	Peaks 峰值	Peaks 峰值
		Gradient potentials 电位梯度	Zeros 零点	Zeros 零点	Zeros 零点
	Mise-a-la-masse (after filtering) 充电法（滤波后）	Potentials 电位	Zeros 零点	Zeros 零点	Zeros 零点
		Gradient potentials 电位梯度	Peaks 峰值	Peaks 峰值	Peaks 峰值
Well 钻孔			SK15	SK14	SK13
Borehole structure 钻孔类型	Borehole depth 钻孔深度 /m		310.7	249.1	251.5
	Opening elevation 孔口高程 /m		1224	1225.6	1230.6
	Opening aperture 孔口孔径 /mm		200	200	200
	Final aperture 终孔孔径 /mm		130	125	130
Water bearing section 含水段	Water table in elevation 水位标高 /m		1070	1070.5	1072.6
	Type 类型		Channels 管道	None 无	Fractures 裂隙带
	Thickness 厚度 /m		60.5	None 无	32
Pumping test 抽水试验	Drawdown 降深 /m		20.5	None 无	56.5
	Flux 流量 / (L/s)		2.82	None 无	1.45
	Recovery time 恢复时间 (h:m)		8:30	None 无	9:30

5. Groundwater occurrence characteristics and drilling well models in karst slope zone, Bijie, Guizhou Province

贵州毕节岩溶斜坡地带地下水赋存规律与钻探成井模式

Combined with the hydrogeological survey data and 486 drilling data obtained in Bijie area of Guizhou Province, the study found that structure, lithology and geomorphology are the main factors controlling boreholes, which should be combined with concrete examples. The Cambrian Loushanguan Formation ($\epsilon_{2-3}ls$), the Permian Qixia–Maokou Formation (P_2q-m), the Triassic Jialingjiang Formation (T_1j), and the Guanling Formation (T_2g) are the main karst water-finding strata, which account for 79.83% of all the wells, and the well completion rates are 88.1%, 41.56%, 76.32% and 70.94%, respectively. The average well water inflows are 409.62 m^3/d, 165.93 m^3/d, 291.2 m^3/d, and 277.42 m^3/d, respectively; compared with the thick layer of pure limestone, the limestone intercalated with classic and dolomite strata have a higher well formation rate, and the average water inflow of the borehole is greater. The regional fault is no longer the favorable part of drilling for water exploration. The small structures such as small and medium faults control the local water-rich area as the favorable target area for drilling water, and this study sums up 4 drilling patterns: the small and medium-sized fault water control mode, the reverse fault upper plate with rich water mode, the aquifer and the aquiclude contact zone with rich water mode, and the dolomite karst fissure with rich water mode (Pan et al., 2018).

结合贵州省毕节地区水文地质调查资料和486口钻井资料发现：构造、岩性和地貌是控制钻孔是否成井的主要因素，应结合具体实例综合分析。寒武系娄山关组（$\epsilon_{2-3}ls$）、二叠系栖霞－茅口组（P_2q-m）、三叠系嘉陵江组（T_1j）和关岭组（T_2g）为主要岩溶找水层位，占成井总数的79.83%，成井率分别为88.1%、41.56%、76.32%、70.94%，平均单井涌水量分别为409.62 m^3/d、165.93 m^3/d、291.2 m^3/d、277.42 m^3/d。相较于厚层纯灰岩，灰岩夹碎屑岩地层和白云岩地层成井率更高，且钻孔平均涌水量更大；区域性断层不再是钻井找水的有利部位，中小型断裂等小构造控制局部富水区成为钻井找水的有利靶区。归纳出4种钻探成井模式：中小型断裂控水模式、逆断层上盘富水模式、含水层与隔水层接触带富水模式、白云岩岩溶裂隙富水模式（潘晓东等，2018）。

Statistical results of well completion and single well discharge in different karst formations
不同岩溶层位成井数和单井涌水量统计结果表

No. 序号	Water-bearing formations 含水岩组	Stratum Code 地层代号	Lithology 岩性特征	Well completion 成井数/个	Total wells 总井数/个	Proportion to total boreholes 占总钻孔比例/%	Well completion rates 成井率/%	Range of maximum water yield 最大涌水量范围/(m³/d)	Average water yield per well 平均单井涌水量/(m³/d)
1	Dengying Fm 灯影组	$Zbdn$	Thin to thick dolomite, clastic dolomite 薄至厚层白云岩、碎屑白云岩	15	17	3.09	100.00	60.3~840.33	289.304
2	Qingxudong Fm 清虚洞组	$\epsilon_1 q$	Medium thick limestone mixed with dolomitic limestone, dolomite 中厚层灰岩夹白云质灰岩、白云岩	5	7	1.44	71.43	0~295.3	122.89
3	Loushanguan Fm 娄山关组	$\epsilon_{2-3} ls$	Thin to massive fine-grained dolomite, argillaceous dolomite 薄层至块状晶白云岩、泥质白云岩	37	42	8.64	88.10	0~1767.9	409.62
4	Baizuo Fm 摆佐组	$C_1 b$	Gray white thick massive dolomite 灰白色厚层块状白云岩	18	21	4.32	85.71	0~944	262.97
5	Datang Fm 大塘组	$C_1 d$	Limestone mixed with shale, sandstone 灰岩夹页岩、砂岩	1	1	0.21	100.00	68.6	68.6
6	Huanglong Fm 黄龙组	$C_2 hn$	Thin to thick dolomite 中厚层白云岩	9	15	3.09	60.00	0~485.43	152.38
7	Maping Fm 马坪组	$C_3 mp$	Medium thick to massive limestone 中厚层至块状灰岩	4	8	1.65	50.00	0~405.28	82.97
8	Qixia–Maokou Fm 栖霞–茅口组	$P_2 q\text{-}m$	Thick to massive limestone 厚层至块状灰岩	32	77	15.84	41.56	0~2863.21	165.93
9	Yelang Fm 夜郎组	$T_1 y$	Upper and Lower: mudstone and siltstone; Middle: limestone 上段和下段为泥岩、粉砂岩，中段为灰岩	21	31	6.38	67.74	0~1328.4	237.04
10	Jialingjiang Fm 嘉陵江组	$T_1 j$	Thin to medium thick limestone, argillaceous limestone mixed with mudstone 薄至中厚层灰岩、泥质灰岩夹泥岩	116	152	31.28	76.32	0~2863.21	291.2
11	Guanling Fm 关岭组	$T_2 g$	Lower: argillaceous limestone, dolomite; Middle: limestone; Upper: dolomite 下段为泥质灰岩、白云，中段为灰，上段为白云岩	83	117	24.07	70.94	0~1431.03	277.42
Total 总计	—	—	—	341	486	100	70.16	0~2863.21	262.94

6. Optimizing the regulation of karst water to enable efficient utilization in Luxi Xiaojiang River Basin
泸西小江流域岩溶水优化调控、高效利用模式

In the Luxi karst faulted basin demonstration area, the study has developed a technical system for karst groundwater regulation and utilization, and set up a mode for joint regulation of water resources in basins. The study developed epikarst spring - pool (cellar) - reservoir combined regulation technology driven by photovoltaic energy in the surrounding mountainous areas, raising the utilization efficiency of groundwater by 28.6%; in the area lack of water with uneven aquifers, the study developed the technology of high-power charging with cross hole CT imaging to locate groundwater accurately and combined the technology of farmland water-saving irrigation and soil moisture conservation to enhance the utilization rate of surface-underground water resources by more than 50%. These technologies helped locals to solve the water problem for production and living effectively. The epikarst spring - pool (cellar) - reservoir combined regulation technology has achieved the most typical efficiency, take the Wabushan Reservoir and Wanbankong Epikarst Spring Project as an example: ① used anti-seepage curtain grouting technology in Wabushan Reservoir to build a suspended reservoir at a karst depression with a storage capacity of 140,000 m^3 to store surface water and a small amount of epikarst water, and build a water supply network for downstream villages; ② built a water pool near the Wanbankong Epikarst Spring, established a water supply network to downstream villages, and exploited the epikarst water with an exploration rate of about 73,000 m^3/a; and ③ established a water pump station and a photovoltaic power station near the Wanbankong Epikarst Spring, use the photovoltaic power station to generate electricity automatically and the pump station to pump water automatically to recharge the Wabushan Reservoir upstream, when epikarst spring has relative great discharge during normal water season by taking advantage of its delayed regulation and storage function to precipitation. The recharge from the pumping water is about 40,000 m^3/a, means that the storage capacity of the Wabushan Reservoir has been expanded from 140,000 m^3 to 180,000 m^3 without expanding the actual storage volume, enhancing the water supply capacity of the reservoir in dry season, and increasing the utilization rate of water resources (water supply capacity of the reservoir) by 28.6% (Lan et al., 2021).

在泸西岩溶断陷盆地示范区研发了岩溶地下水调控、利用技术体系，构建了流域尺度水资源联合调控模式。在周边山区研发以光伏为能源驱动的表层泉－池（窖）－库联合调控技术，地下水资源利用效率提高到28.6%；在盆地缺水、含水介质不均的地段，研发大功率充电联合跨孔CT成像准确定位地下水技术，并结合农田节水灌溉与土壤保墒等技术的研发，使地表－地下水资源利用率提高50%以上。这些有效解决了当地生产、生活用水问题。其中表层泉－池（窖）－库联合调控技术成效最为显著，以泸西凹部山水库、湾半孔表层泉开发示范工程为例：①在凹部山采用防渗帷幕灌浆技术，建成悬挂式的凹部山溶洼水库，库容14万m^3，积蓄地表水、少量表层岩溶水，建设供水管网供给下游村寨用水；②在湾半孔表层泉附近建设蓄水池，建设供水管网，供下游村寨用水，开采表层岩溶水，开采量约7.3万m^3/a；③在湾半孔表层泉附近建设水泵站、光伏电站，利用表层岩溶水对降雨的延迟调蓄功能，在平水期表层泉流量还较大时，通过光伏电站自动发电，泵站自动抽取表层岩溶水补给上游凹部山水库，抽水补水量约4万m^3/a，相当于在无法扩大水库实际库容的前提下，将凹部山水库库容由14万m^{3}扩大到18万m^3，直接扩大水库旱季供水能力，使水资源利用率（水库供水能力）提高到28.6%（蓝芙宁等，2021）。

Chapter 3 Scientific Research

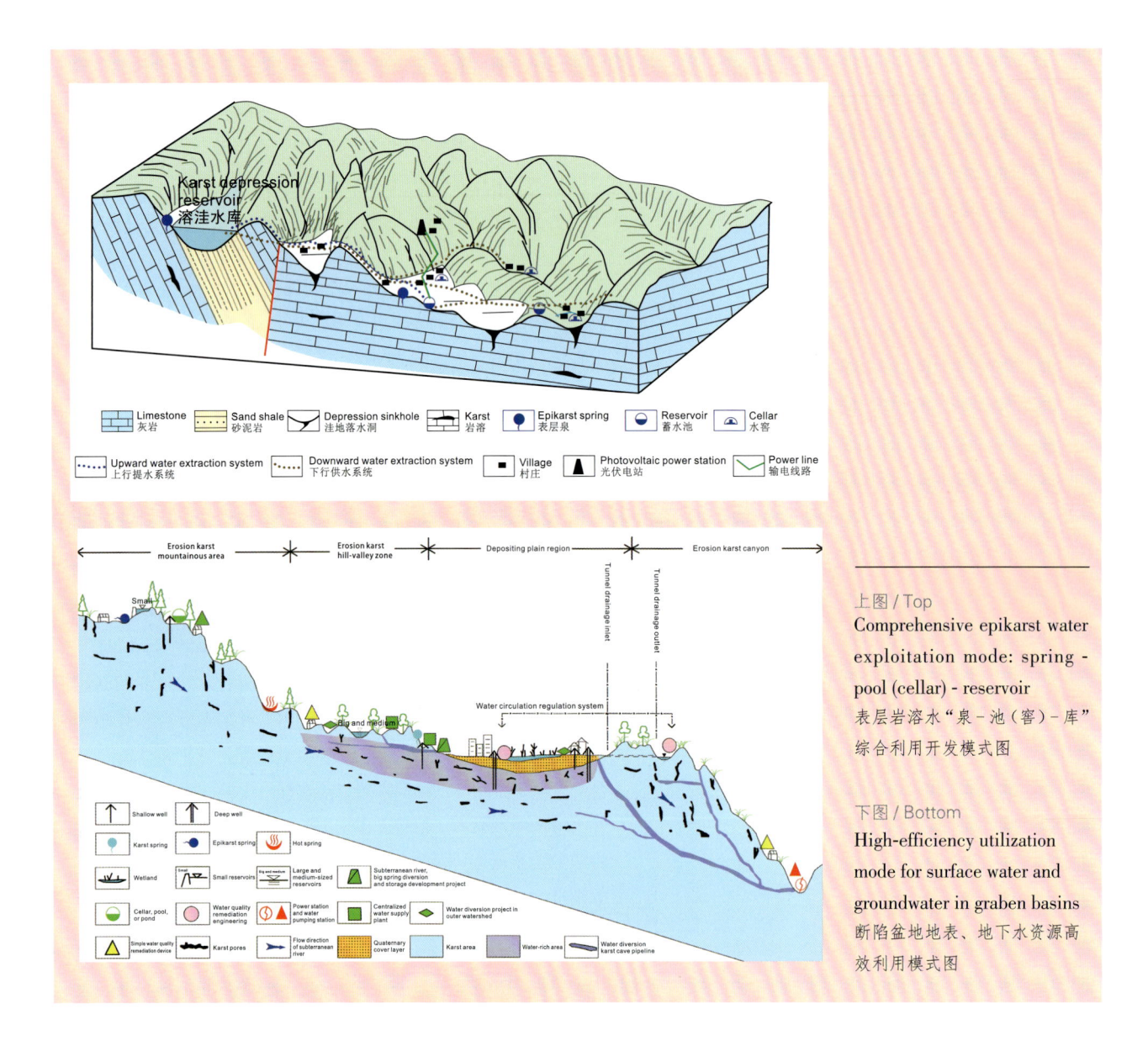

上图 / Top
Comprehensive epikarst water exploitation mode: spring - pool (cellar) - reservoir
表层岩溶水"泉－池（窖）－库"综合利用开发模式图

下图 / Bottom
High-efficiency utilization mode for surface water and groundwater in graben basins
断陷盆地地表、地下水资源高效利用模式图

7. The research achievements and thoughts of karst water in northern China

中国北方岩溶水研究进展与方向思考

The carbonate aquifer in northern China is an important water resource of the country with large thickness and wide distribution. The karst water in this vast region has the functions in water resources, tourism, and ecology. It plays an important role in the construction of national economy, especially in the construction of urban water supply and energy infrastructures. In the past 50 years, due to the influence of large-scale development and other high-intensity human activities, karst water environment problems in northern China have become prominent and brought a series of negative effects, which seriously restrict the sustainable utilization of water resources. The study takes the Jinci Spring as an example and proposes the reflow measures: ① raising the water level of Reservoir-II to the designed elevation of 905.7 m that can increase the leakage of 5.84×10^4 m³/d; ② closing the depressurization drainage well of Baijiazhuang Coal Mine (closed already) in the strong runoff zone of karst groundwater, reducing the discharge of 0.8×10^4 m³/d; ③ treating the dewatering of some artesian wells at the downstream of Jinci Spring outlet and Dongyu Coal Mine, which may decrease the discharge from 2.3×10^4 m³/d to 1.15×10^4 m³/d. Based on the results from the numerical model of groundwater seepage, it is possible to realize the reflow of Jinci Spring in 2–3 years without a great increase of karst water exploitation or suffering from extreme droughts by the abovementioned measures. The study takes "old kiln water" of Shandi River Coal Mine as an example and proposes four measures: ① closing open-pit mines (Xiaogou Mine and Zhuangzhi Mine) physically; ② covering top soil and vegetation on uncovered gangue and slag; ③ treating old kiln water in tunnels, carrying out neutralization+microbial treatment or direct microbial treatment according to oxidation or reduction water logging areas, and carrying out in situ pilot treatment based on microbial research foundation; ④ treating the treated "old kiln water" at the end collectively may reduce the cost and decrease the treatment on sediments. Moreover, the study thinks that there is great exploitation potential for hypogene karst water and geothermal resources, and puts forward that the following areas could be targets for exploration and exploitation in priority: the piedmont of Taihang Mountain central-south section, the piedmont of Taihang Mountain north section, the Jibei and Qinshui syncline north edge in Shandong, the east edge of Ordos Basin and Fen-Wei Graben (rift basin); the upper Cambrian Fengshan Formation coarse dolomite is another targeted stratum except for the Middle Ordovician and Jixianian System, among the carbonate reservoirs for water and geothermal resources in northern China (Liang et al., 2021).

中国北方碳酸盐岩含水层连续沉积厚度大、分布范围广，蕴藏着丰富的地下水资源，是重要的国家级含水层。北方岩溶水集水资源、旅游资源、生态功能等于一体，在国民经济建设中具有举足轻重的地位，特别是在城市供水和能源基地建设中发挥了不可替代的支撑性作用。近50年来，受大规模水资源开发利用和人类其他高强度活动影响，北方岩溶水环境问题凸显并带来了一系列负面效应，严重制约了水资源的可持续利用。研究以晋祠泉水复流为例，提出了晋祠泉的复流措施：①抬高二库蓄水水位至设计标高905.7 m，可增加渗漏量5.84万 m^3/d；②关闭处于岩溶地下水强径流带内的白家庄煤矿（已闭坑）降压排水井，减排0.8万 m^3/d；③处置晋祠泉口下游部分自流井及东于煤矿的排水，可使总排泄量从现状2.3万 m^3/d 减少到1.15万 m^3/d。根据已建立的泉域地下水渗流数值模型预测结果，实施上述3项近期措施，在不大量增加泉域岩溶水开发和发生极端干旱气候条件下，可望于2~3年实现晋祠泉水复流。以山底河煤矿"老窑水"治理为例，提出4项治理措施：①物理封闭露天矿（小沟露天矿、庄只露天矿）；②裸露煤矸石及矿渣分布区的浮土覆盖与绿化；③坑道内处理，根据氧化、还原积水区分别开展中和+微生物、直接采用微生物法治理，鉴于微生物法的研究基础，需要开展原位中试治理；④对经源头治理后的"老窑水"出流开展集中末端处理，可降低成本并减少对沉淀物的处置。此外，本研究认为北方岩溶区深部岩溶水、热资源开发潜力巨大，提出将山盆接合部位的太行山中南段山前、太行山北段山前、山东济北和沁水向斜北缘、鄂尔多斯盆地东缘及裂谷型盆地的汾渭地堑作为靶区，优先进行勘探开发；在北方具有水热资源的碳酸盐岩储层中，除了中奥陶统和蓟县系外，上寒武统凤山组粗晶白云岩是值得关注的目标层位（梁永平等，2021）。

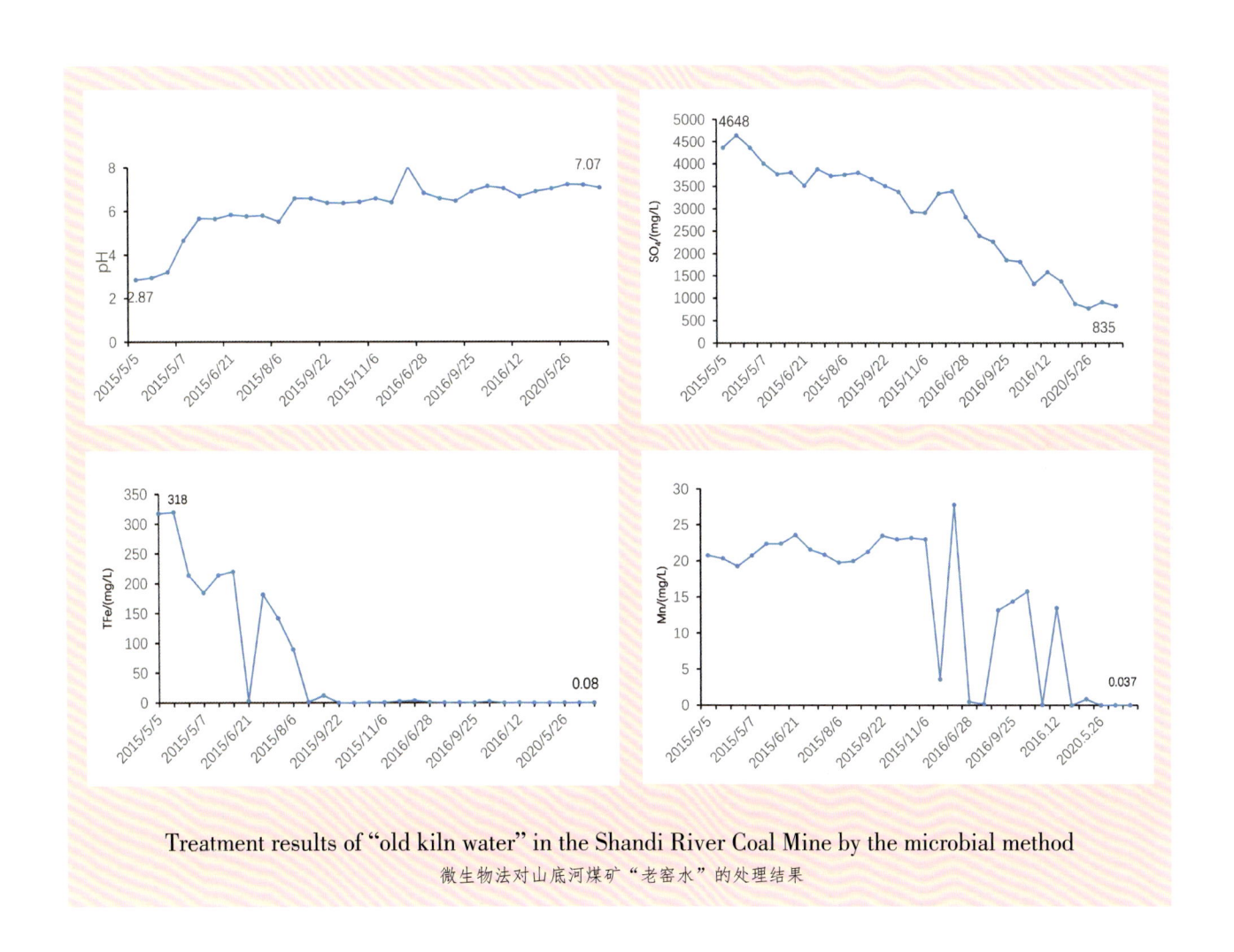

Treatment results of "old kiln water" in the Shandi River Coal Mine by the microbial method

微生物法对山底河煤矿"老窑水"的处理结果

8. Contamination characteristics of chlorinated hydrocarbons in a fractured karst aquifer using TMVOC and hydro-chemical techniques

基于 TMVOC 和水化学技术的岩溶裂隙含水层中氯代烃的污染特征

The study investigated a fractured karst aquifer polluted by chlorinated hydrocarbons to determine the contamination characteristics of the main hydrocarbon components. The natural attenuation processes of representative components were simulated and forecasted using TMVOC and hydro-chemical components (NO_3^-, SO_4^{2-}, HCO_3^-, Cl^-, and $\delta^{13}C_{DIC}$). The impacts of hydrocarbon compounds on the hydro-chemical ions were estimated, and their historical contamination characteristics were also reconstructed. Results showed that the dynamic characteristics of Trichloromethane and 1,1,2-Trichloroethane can indicate those of chlorinated hydrocarbons, where the rate of natural attenuation was observed to decrease with decreasing concentrations of hydrocarbon compounds. Additionally, the long-term variation characteristics in groundwater levels showed that the relatively stable hydrodynamic field conditions enabled the simulation of the natural attenuation processes of chlorinated hydrocarbons. The simulation which also considered the biodegradation processes showed that the use of TMVOC and hydro-chemical parameters may better describe natural attenuation processes. Over 3 years (from 2017 to 2019), the average percentage of biodegradation in the total natural attenuation was estimated to be 88.35%. Similarly, Trichloromethane and 1,1,2-Trichloroethane are forecasted to have no health hazards in 10 and 15 years, respectively. The contribution rates of biodegradation to HCO_3^- and Cl^- in the fractured karst aquifer varied with the concentrations of chlorinated hydrocarbons. Overall, the findings and methods in this work have significant contributions to advancing remediation developments of petroleum hydrocarbons, especially in karst environments that are highly susceptible to contamination (Guo et al., 2021).

本研究对一个受氯代烃有机物污染的岩溶裂隙含水层开展调查，以确定其主要烃组分的污染特征。利用 TMVOC 和水化学组分（NO_3^-、SO_4^{2-}、HCO_3^-、Cl^- 和 $\delta^{13}C_{DIC}$）模拟和预测代表性组分的自然衰减过程，估算碳氢化合物对水化学离子的影响，并反演出其历史污染特征。研究结果表明，三氯甲烷和 1,1,2- 三氯乙烷的动态特征可指示氯代烃有机物的动态特征，其自然衰减速率随着碳氢化合物浓度的降低而降低。此外，区内地下水位的长期变化特征表明，相对稳定的水动力场条件有利于模拟氯代烃有机物的自然衰减过程。考虑生物降解过程的模拟表明，TMVOC 和水化学参数的使用可更好地表征氯代烃有机物的自然衰减过程。在 3 年（2017~2019 年）内，生物降解量占自然衰减总量的平均百分比为 88.35%。三氯甲烷和 1,1,2- 三氯乙烷预计在 10 年和 15 年内不会对健康造成危害。岩溶裂隙含水层中氯代烃的生物降解作用对 HCO_3^- 和 Cl^- 浓度的贡献率随氯代烃有机物浓度的变化而变化。总的来说，本次研究成果和方法有利于推动石油碳氢化合物的修复技术的研发，特别是在极易受污染的岩溶环境中（Guo et al.，2021）。

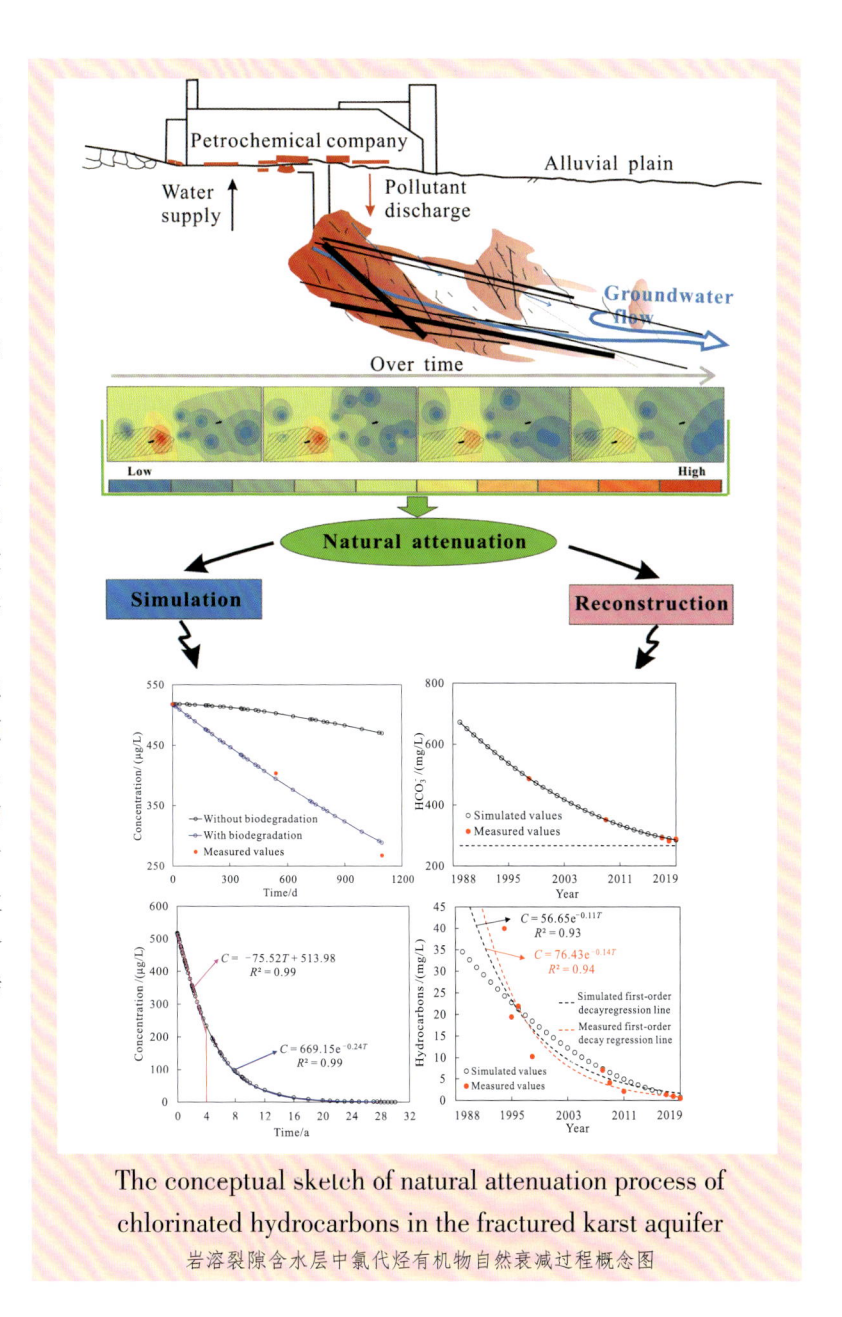

The conceptual sketch of natural attenuation process of chlorinated hydrocarbons in the fractured karst aquifer
岩溶裂隙含水层中氯代烃有机物自然衰减过程概念图

9. Groundwater resources evaluation in the Pearl River Basin based on SWAT model

基于 SWAT 模型的珠江流域地下水资源评价研究

On the basis of combing the current situation and history of groundwater resources evaluation in the Pearl River Basin, the study discusses the basic principle and basic database of SWAT distributed hydrological model. The research divided the Pearl River Basin into 129 four-level groundwater systems, and fully considered the parameter sensitivity of karst, bedrock fissures, and pore aquifer media. Based on the monthly measured runoff of 9 hydrological stations from 2008 to 2016, carried out parameter calibration and model calibration, calculated the rainfall assurance rate in different years according to the rainfall from 1957 to 2017, analyzed and evaluated the multi-year groundwater resources in the Pearl River Basin. Finally, the study carried out the inversion of rainfall infiltration coefficient and groundwater recharge model parameters. Through this assessment, the average recharge of the Pearl River Basin from 2010 to 2016 is 148.802 billion m^3, and the total recharge of groundwater in extra dry years (2011), normal years (2010), and high water years (2016) are 71.949 billion m^3, 144.682 billion m^3, and 178.187 billion m^3, respectively, of which the rainfall in high water years is about 1.7 times that in extra dry years, the groundwater recharge is 2.48 times. The total annual recharge of bedrock fissure and pore water bearing medium is 52.991 billion m^3, 44.513 billion m^3, and 51.298 billion m^3, respectively. Through parameter inversion, the average annual groundwater recharge modulus is 10.83 $L/(s·km^2)$ and the rainfall infiltration coefficient is 0.246, in order to provide data support and scientific services for groundwater development, utilization, treatment and protection in the basin (Zhao et al., 2022).

在梳理珠江流域地下水资源评价现状及历史的基础上，讨论了 SWAT 分布式水文模型基本原理和基础数据库，将珠江流域划分为 129 个地下水子流域，在充分考虑岩溶、基岩裂隙及孔隙含水介质参数敏感性的基础上，基于 9 个水文站 2008~2016 年逐月实测径流量进行参数率定和模型校准，并根据1957~2017年的降水量计算不同年份的降雨保证率，分析评价珠江流域多年地下水资源量，最后开展降雨入渗系数和地下水补给模数参数反演。通过本次评价，得到珠江流域2010~2016年平均补给量为1488.02亿m³，特枯年（2011年）、平水年（2010年）及丰水年（2016年）地下水总补给量分别为719.49亿m³、1446.82亿m³、1781.87亿m³，其中丰水年的降水量约为特枯年的1.7倍，地下水补给量为2.48倍，岩溶、基岩裂隙及孔隙含水介质的年均补给总量分别为529.91亿m³、445.13亿m³、512.98亿m³，通过参数反演获得年均地下水补给模数为10.83 L/(s·km²)，降雨入渗系数为0.246，以期为流域内地下水开发利用与治理保护提供数据支撑和科学服务（赵良杰等，2022）。

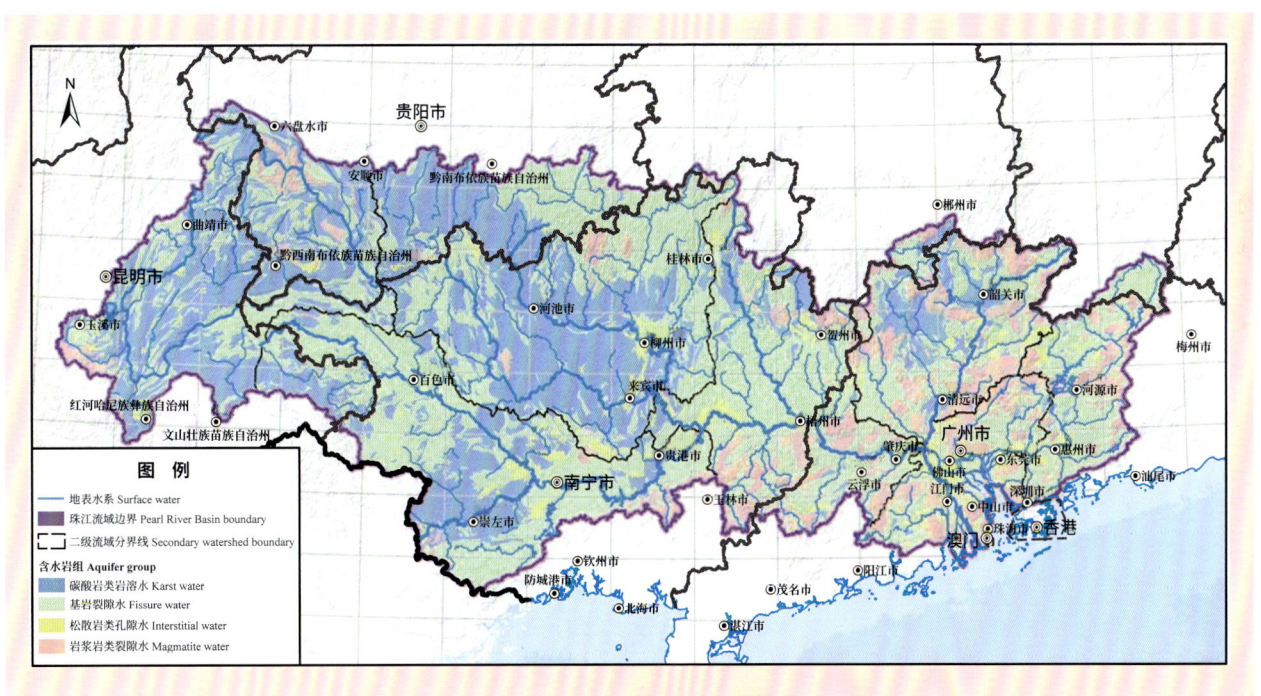

Water-bearing rock formations and secondary watershed boundary in the Pearl River Basin
珠江流域含水岩组及二级分区

10. The study of the solute transport model for karst conduits based on CFP
基于 CFP 的岩溶管道流溶质运移数值模拟研究

To study the characteristics of solute transport for karst conduits, a coupling method of conduit flow process (CFP) and the modular three-dimentional transport model (MT3DMS) is proposed. For the purpose of probing into the impacts of hydrogeological parameter on solute transport for karst conduits, based on analysis of the fundamentals of CFP and MT3DMS, the research established the concept model with one sink hole and four conduits. The results of the coupling model show that the CFP model permits the exchange of groundwater between matrix and pipe, and the tailing feature of the breakthrough curve for karst conduits is well depicted. With the decreasing effect from porosity and the increasing effect from pipe diameter and the increasing pipe conductance and hydraulic gradient, the curve peak became larger, and the time to reach the peak becomes quicker and the symmetry of the curve is more apparent. Thus, the coupling model of CFP and MT3DMS can depict the characteristics of solute transport for karst conduits. How to generalize the physical characteristics of karst conduits is a way to resolve the multi-peak issues (Yang et al., 2019).

多重岩溶含水介质的复杂性导致岩溶地下水流动及溶质运移的数学模拟成为地下水研究难点之一。为了探讨岩溶多重含水介质中地下水流溶质运移特征，本研究构建了管道流 CFP 水流模型和 MT3DMS 溶质运移三维耦合数值模型。在阐述管道流 CFP 和 MT3DMS 基本原理的基础上，通过建立水文地质概念模型算例(1个落水洞、4个直管道)，探讨岩溶管道水流及溶质运移规律，分析讨论不同水文地质参数对浓度穿透曲线的影响。研究结果表明：管道流 CFP 模型能够刻画岩溶管道与基岩裂隙水流交换特征，MT3DMS 模型能够模拟穿透曲线的拖尾现象，符合实际岩溶区特征。随着水力梯度、管道直径及管道渗透系数增大，孔隙度减小，浓度曲线峰值越大，峰值到达时间越快，浓度穿透曲线越对称。由此得出结论：耦合 CFP 水流模型和 MT3DMS 溶质运移模型能够刻画岩溶管道流溶质运移规律，为研究岩溶复杂介质污染物运移特征提供了一种思路和途径(杨杨等，2019)。

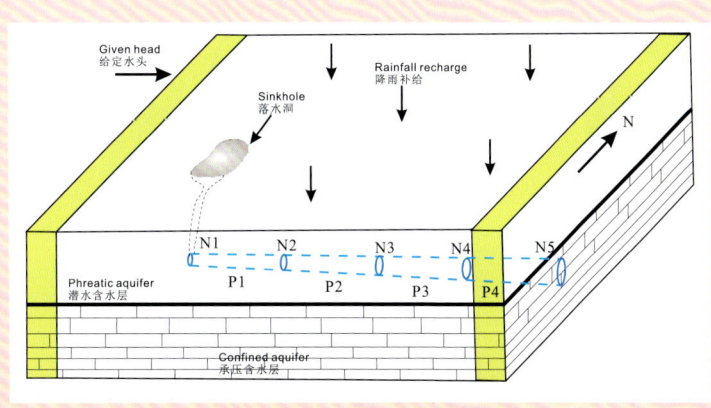

Conceptual model for CFP
管道流 CFP 概念模型

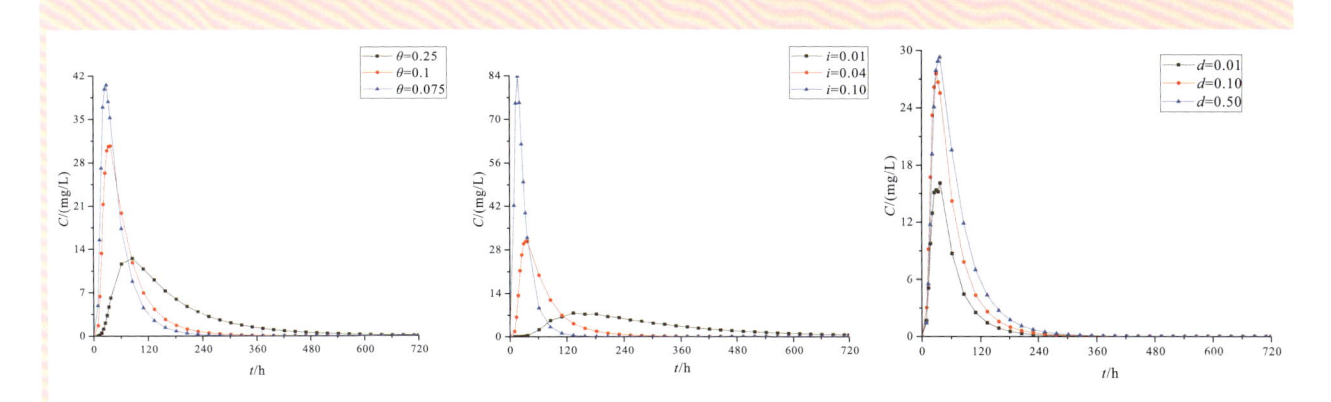

Concentration curves with different porosities (left), curves with different hydraulic gradients (middle), curves with different conduit diameters (right)
不同孔隙度接收点浓度曲线（左）、不同水力梯度接收点浓度曲线（中）、不同管道直径接收点浓度曲线（右）

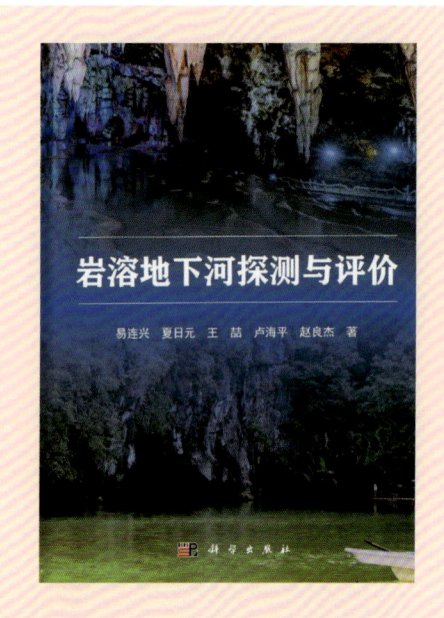

上左图 / Top left
Investigation, Evaluation, Development and Utilization Mode of Groundwater Resources in Karst Mountainous Areas in Southwest China
《西南岩溶石山区地下水资源调查评价与开发利用模式》

上右图 / Top right
Detection and Evaluation of Karst Underground Rivers
《岩溶地下河探测与评价》

下图 / Bottom
Study on Water Cycle Mechanism of Typical Karst Underground River Systems
《典型岩溶地下河系统水循环机理研究》

3.2.3 Karst ecosystem and ecological restoration
岩溶生态系统与生态修复

1. Evolution features of rocky desertification and influence factors in karst areas of southwest China in the 21st century
21世纪西南岩溶石漠化演变特点及影响因素

By analyzing the remote sensing results and relevant statistical data of the rocky desertification area in the southwest karst area in China, the spatiotemporal evolution features and influencing factors of rocky desertification since the 21st century have been revealed. In 2015, the total area of rocky desertification in the karst areas of southwest China decreased to 9.2×10^4 km^2, with the overall trend of rocky desertification evolution changed from intensified before the 21st century to gradually alleviated in the 21st century. Moreover, the degree of rocky desertification in the karst areas of southwest China has significantly decreased, from severe to moderate rocky desertification at the beginning of the 21st century to light to moderate rocky desertification. The proportion of the most harmful severe rocky desertification has decreased from 38.08% to 15.31%, indicating an effective curb on the trend of rocky desertification. However, there are significant regional differences in the evolution of rocky desertification, which are mainly related to the national implementation of vegetation restoration projects, the types of karst landforms that affect vegetation restoration, the degree of groundwater development, rainwater resources, and regional economic conditions. The planned area of vegetation construction is directly proportional to the reduced area of rocky desertification. Peak forest plains and karst depressions with relatively good ecological and economic conditions have the best effect in controlling rocky desertification. Groundwater development and relatively abundant rainfall can effectively promote vegetation restoration and economic development, and the poverty of residents may worsen rocky desertification (Jiang et al., 2016).

通过分析我国西南岩溶地区石漠化面积遥感调查结果和相关统计资料，揭示了21世纪以来石漠化时空演变特征和影响因素。2015 年，我国西南岩溶地区石漠化总面积降至 9.2 万 km²，石漠化演变的总趋势由 21 世纪以前的加剧变化为 21 世纪的逐渐减缓。而且，西南岩溶区石漠化程度显著变轻，由 21 世纪初的以重、中度石漠化为主演变为以轻、中度石漠化为主，危害最大的重度石漠化面积比例由 38.08％降至 15.31％，说明石漠化趋势得到有效遏制。但石漠化演变存在较大的区域差异，主要与国家实施植被修复工程的力度、影响植被恢复的岩溶地貌类型、地下水开发程度、雨水资源及区域经济条件密切相关。植被建设规划面积与石漠化减少面积成正比，生态经济条件相对较好的峰林平原和溶丘洼地石漠化治理效果最好，地下水开发和比较丰沛的降水可有力促进植被恢复和经济发展，而居民贫困化可使石漠化恶化 (蒋忠诚等，2016)。

Statistics of rocky desertification area changes in 8 provinces (regions or cities) of southwest China from 2000–2015
西南八省（区、市）岩溶石漠化演变情况统计表 (2000~2015 年)

Province 省份	Karst area 岩溶区面积 / 10⁴ km²	Rocky desertification degree in 2000 2000 年石漠化程度比例 /% Severe 重度	Moderate 中度	Mild 轻度	Rocky desertification degree in 2015 2015 年石漠化程度比例 /% Severe 重度	Moderate 中度	Mild 轻度
Yunnan 云南	10.86	28.23	36.38	35.39	26.60	35.21	38.18
Guizhou 贵州	12.12	47.21	36.63	16.16	9.52	35.05	55.43
Guangxi 广西	8.17	32.65	42.15	25.20	13.84	29.74	56.42
Sichuan 四川	6.91	37.90	20.93	41.17	14.18	40.15	45.67
Hunan 湖南	6.25	52.10	32.99	14.91	0.28	34.58	65.14
Hubei 湖北	5.04	46.54	37.00	16.46	9.03	34.66	56.31
Chongqing 重庆	3.01	49.97	32.52	17.50	6.05	35.64	58.31
Guangdong 广东	1.19	45.80	33.33	20.87	12.48	38.39	49.13
Total 总计	53.56	38.08	36.92	25.00	15.31	34.24	50.45

2. Serial outcomes on controlling rocky desertification in karst faulted basin
岩溶断陷盆地石漠化治理系列成果

1) Rocky desertification evolution, treatment and demonstration in karst faulted basins, Southwest China
中国西南岩溶断陷盆地石漠化演变及治理技术与示范

Karst faulted basin area is the area for national integrated rocky desertification control, ecological security shelter, and contiguous poverty-stricken area, but with less science and technology investment. Aiming to solve problems like the environmental geological structural differentiation of coexistence basin and mountain, as well as unmatched water and soil resources and severe rocky desertification problems, it is planned to implement the National Key Research & Development Program of China "Rocky desertification evolution, integrated control technologies and demonstration in karst faulted basin area, Southwest China" (2016YFC0502500) in Mengzi, Jianshui, and Luxi of eastern Yunnan. The Program is hoping to illustrate the rocky desertification evolution mechanism at catchment scales, develop the integrated technology for rocky desertification with the highly efficient utilization of water and soil resources as the basis and the enhancement of ecological service function as the major focus, formulate a collaborative ecological treatment - ecological industry development mode, so as to provide technology and demonstration for ecological enriching and ecological civilization construction. The Program hopes to generate serial outcomes on identifying rocky desertification in karst faulted basin, developing treating technology, establishing ecological industry, evaluating comprehensive benefits, and proposing measures for highly efficient space use of land resources (Cao et al., 2016).

岩溶断陷盆地区是国家石漠化综合治理工程示范区、生态安全屏障区和连片特困区，同时也是石漠化治理科技投入薄弱区。针对断陷盆地盆-山共存的环境地质结构分异及水土资源不匹配和石漠化严重等问题，以滇东蒙自、建水、泸西为重点区，开展"喀斯特断陷盆地石漠化演变及综合治理技术与示范"(2016YFC0502500)项目的国家重点研发活动，阐明流域尺度石漠化演变机理，以水土资源高效利用为基础、生态服务功能提升为核心，研发石漠化综合治理技术，形成生

态治理 – 生态产业协同发展模式，为生态富民和生态文明建设提供技术与示范。项目希望形成岩溶断陷盆地石漠化机理识别 – 治理技术研发 – 生态产业模式构建 – 综合效益评估 – 国土资源空间高效利用对策等一系列成果（曹建华等，2016）。

2) Zoning of environmental geology and functions in karst faulted basins
岩溶断陷盆地环境地质分区及功能

Karst faulted basin is dominated by local fault blocks subsidence that is accompanied by erosion and dissolution. The basin is featured by distinct geomorphic zoning, intense neotectonic movements, uneven water resources, prominent zoning of soil and vegetation, significant vertical variations of climate, and great regional differences of human activities. Among them, the types and causes of the geomorphic landscapes are not only the overall reflection of the environmental geology but also the influencing factors of the habitat. Moreover, the types and causes of the landscapes are outstanding features for environmental geological zoning, as they can be easily observed and distinguished. Accordingly, the karst faulted basins can be divided into four environmental geological zones: the erosion karst mountainous area, the erosion karst hill-valley zone, depositing plain region, and the erosion karst canyon. Each of these zones has different functions: the erosion karst mountainous area has highly significant ecological functions; the erosion karst hill-valley zone has equally important ecological and economic functions; the depositing plain region has the most remarkable economic functions; and the erosion karst canyon has the functions decided by its location of the associate river systems and its macro influence. Karst faulted basin has low susceptibility of geological disasters. Although karst distributed widely, yet there is few karst collapses, showing point-like distribution characteristics. The resource development and environment recovery of karst faulted basins should be coordinated with environmental geology. It is necessary to evaluate the resources and environment based on catchments and weigh the advantages and disadvantages generally with scientific planning to achieve the sustainable effects (Wang et al., 2017).

岩溶断陷盆地以局部地块断陷为主导并伴随侵蚀及溶蚀，具有地貌分区明显、新构造运动强烈、水资源分布不均、土壤和植被分带性强、气候垂向变化明显、人类活动区间差异大等特征。其中，地貌形态和成因类型既是环境地质本质特征的综合反映，又是对生境具有重大影响的自然因素，而且容易认识和识别，是环境地质分区的良好标志。以地貌形态和成因类型为标志，岩溶断陷盆地可划分为侵溶蚀山区、溶蚀丘峰谷地区、沉积平坝区、侵溶蚀河谷区4个环境地质分区。侵溶蚀山区生态功能突出，溶蚀丘峰谷地区生态与经济功能并重，沉积平坝区经济功能尤显重要，侵蚀河谷区视所处大江大河流域的区位及宏观影响而定。岩溶断陷盆地地质灾害易发性低，虽然岩溶分布面积广大，但岩溶塌陷较少，呈零星"点状"分布特征。岩溶断陷盆地的资源开发和环境恢复治理，应与环境地质分区功能相协调，按流域系统评价并全面权衡利弊、科学规划布局，才能取得可持续的成效（王宇等，2017）。

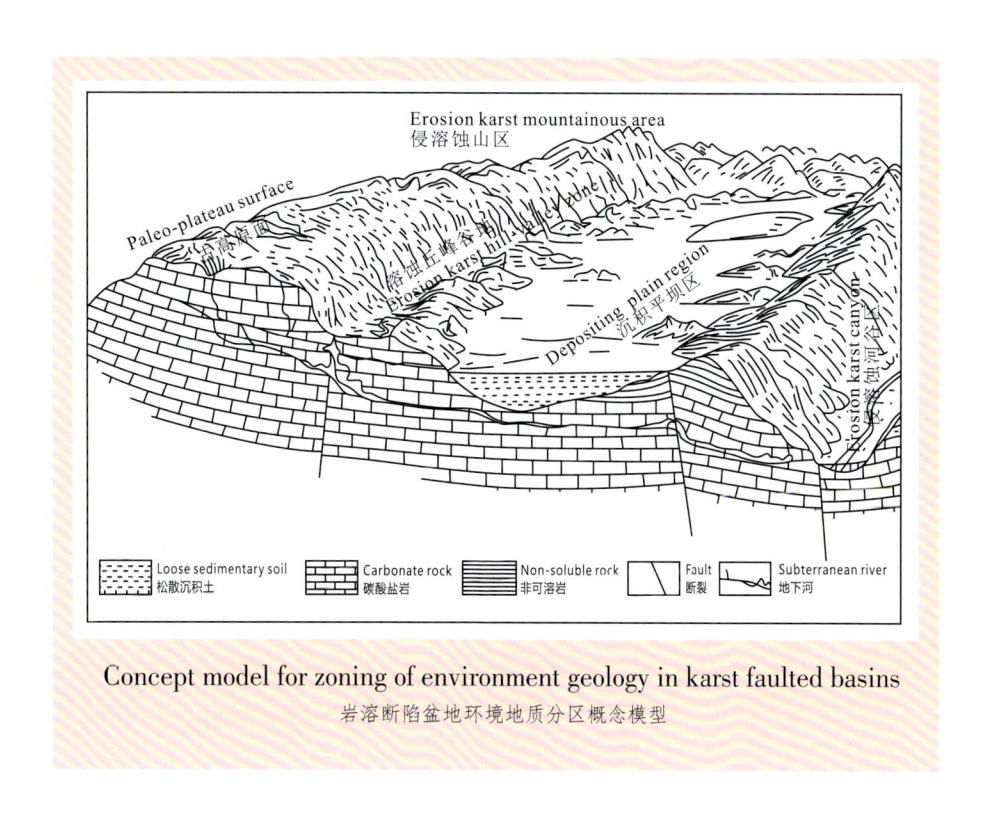

Concept model for zoning of environment geology in karst faulted basins
岩溶断陷盆地环境地质分区概念模型

3) Climatic characteristics under the influence of basin-mountain coupled topography and its influence on the ecological restoration of rocky desertification in Mengzi karst fautled basin, Southwest China
中国西南蒙自岩溶断陷盆地"盆－山"耦合地形影响下的气候特征及其对石漠化生态恢复的影响

The study has set up three meteorological observation stations in the basin, on the slope and mountainous region in Mengzi faulted basin in Yunnan Province to collect the meteorological data of the vertical profiles. The vertical climate characteristics and the possible influence on the evolution of rocky desertification were analyzed on monthly and diurnal scales. The study has obtained the following results: ① The observation section is the leeward slope for the wind mainly comes from the southeast. As a consequence, the annual rainfall is the highest in the mountainous area (1,027.4 mm), followed by that in the basin (662.6 mm), and the lowest on the hillside (574.4 mm). The "valley wind" effect is remarkable, and the valley wind blows during the day, making more rainfall. The topographic fluctuation makes the annual variation coefficient of rainfall in the basin reached 152.36%, which is much greater than that of the hillside (113.81%) and the mountain area (99.36%) and amplified the vertical difference of "dry and wet". The aridity index follows the sequence as the highest in the basin (1.74), followed by that of the hillside (1.70), and the lowest in the mountainous area (0.88). ② The precipitation difference in the vertical direction makes the annual solar radiation in the mountainous area (5,492 MJ/m^2) slightly lower than that of the basin (5,817 MJ/m^2). From the basin to the mountainous area, the vertical temperature gradient is 0.74 ℃/100 m, indicating obvious vertical light and heat differences. ③ The impacts of vertical climate characteristics on the ecological restoration of rocky desertification are as follows: concentrated rainfall combined with steep terrain is easy to accelerate soil erosion; less rainfall during the daytime may enable strong evaporation that might aggravate the loss of soil moisture and hinder the restoration of vegetation; drought-tolerant crops are recommended for vegetation restoration, while photothermal crops are recommended to be planted in basins and thermotropic crops in high-altitude areas by considering the vertical light and heat differences (Wang et al., 2019).

本研究通过在云南蒙自断陷盆地，沿盆地、坡地到山区设立三个小型气象观测站获取的山地垂直剖面的气象数据，从月（季）、日尺度分析石漠化山区"盆－山"耦合地形的气候垂直特征及其对石漠化生态恢复的可能影响。结果表明：①观测剖面是当地主导风东南风的背风坡，年降水量高原面 (1027.4 mm) > 盆地 (662.6 mm) > 坡面 (574.4 mm)；且"山谷风"效应显著，白天吹谷风，降雨比例更大。地形起伏使盆地降雨年变异系数达 152.36%，远大于坡面 (113.81%) 与高原面 (99.36%)，地形放大了垂直方向的"干湿"差异，区域干燥指数盆地 (1.74) > 坡面 (1.70) > 高原面 (0.88)；②垂直方向水汽差异使高原面年太阳辐射量 (5492 MJ/m^2) 略小于盆地 (5817 MJ/m^2)；同时盆地与高原面气温垂直梯度达 0.74℃/100 m，因此在光热条件上存在明显的垂直差异；③垂直气候特征对石漠化生态恢复的影响具体表现在：年内降雨集中，坡度较陡的地形易加速水土流失；降水少，且集中在日间，强烈的蒸发易加剧土壤水分亏缺，不利于植被恢复；研究区水分缺乏，因此在植被恢复治理中应选择耐旱的作物，同时要考虑垂直方向的光热条件差异，盆地选择喜光热的作物，海拔高的地区选择喜温凉的作物 (王赛男等，2019)。

4) The characteristics of soil C, N, and P stoichiometric rations as affected by geological background in karst faulted basin
岩溶断陷盆地地质背景影响下土壤碳、氮、磷元素化学计量比例特征

Soil samples with different degrees of rocky desertification were collected along the basin, slope, and plateau in Mengzi faulted basin of Yunnan Province to study the effects of topography and degree of rocky desertification on soil carbon, nitrogen, and phosphorus stoichiometry. The results showed that under the same topographic conditions, the soil C/N ratio was not affected by the degree of rocky desertification, and was at a relatively stable level, while the C/P ratio and N/P ratio increased with the aggravation of rocky desertification, which was consistent with the result that the content of soil organic carbon and total nitrogen increased with the aggravation of rocky desertification and the phosphorus was relatively stable. Compared with the national average soil level, the C/P ratio and N/P ratio of the three terrain parts in the faulted basin are small, while the C/P ratio and N/P ratio are 3–4 times higher than the national level, indicating that the soil available state content in the faulted basin is low, which is an important limiting factor affecting the growth of vegetation. In addition, the study also found that calcium and pH are important factors affecting the stoichiometry of soil carbon, nitrogen, and phosphorus, and the effect of calcium is much higher than that of pH. In the process of ecological restoration of rocky desertification in faulted basins, not only the altitude and vertical climate factors, but also the tolerance of species to calcium should be considered, and species should be selected according to local conditions (Yang et al., 2019b).

在云南蒙自断陷盆地区沿盆地、斜坡和高原面采集不同石漠化发生程度的土壤，研究地形和石漠化程度对土壤碳、氮和磷元素化学计量的影响。结果发现，同一地形条件下土壤碳氮比不受石漠化程度影响，处于较为稳定水平，而碳磷比和氮磷比则随石漠化程度加剧而增大，这与土壤有机碳和全氮含量随石漠化程度加剧而提高且磷相对稳定的结果一致。与全国土壤平均水平相比，断陷盆地三种地形部位碳磷比和氮磷比较小，而碳有效磷比和氮有效磷比高于全国水平3~4倍，表明断陷盆地土壤有效态元素含量较低，是影响植被生长的重要限制因子。此外，研究还发现钙和pH是影响土壤碳、氮和磷化学计量的重要因子，且钙的影响远高于pH。在断陷盆地石漠化生态恢复时，不仅要考虑海拔和垂直气候因素，还要考虑物种对钙的耐受性，因地制宜选择物种 (Yang et al., 2019b)。

Partial correlation between soil geochemical variables and C, N, and P stoichiometry

土壤地球化学变量与C、N和P化学计量偏相关分析

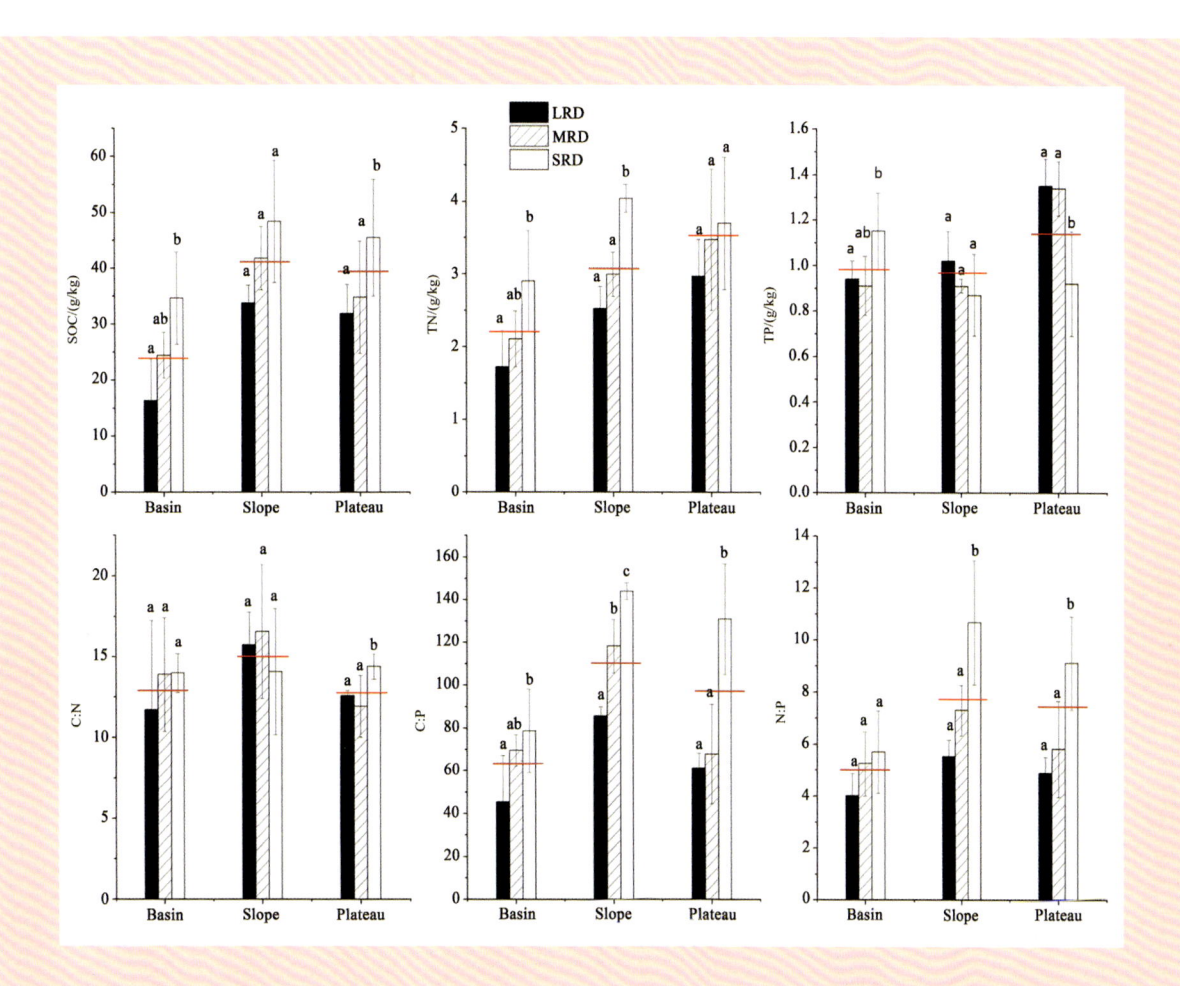

Stoichiometric characteristics of C, N, and P in soils with different rocky desertification levels in different geomorphic parts of faulted basin (LRD: light rocky desertification; MRD: moderate rocky desertification; SRD: severe rocky desertification)

断陷盆地不同地貌部位不同石漠化水平土壤 C、N 和 P 化学计量特征（LRD 为轻度石漠化；MRD 为中度石漠化；SRD 为重度石漠化）

5) Ecological water requirements of crops in typical karst faulted basins: A case study of the Mengzi area, Yunnan Province

典型岩溶断陷盆地农作物生态需水研究——以云南蒙自地区为例

From the perspective of rational regulation of ecological water use, this study explore the ecological water requirements of crops in typical karst faulted basin areas. The study estimates the reference evapotranspiration, ecological water requirements, and the artificial irrigation water requirements of different crops by using the Penman-Monteith formula, crop coefficients (FAO recommended), and the effective rainfall during the same period at three observation sites (Dawazi, Duogu, and Niuerpo) at basin, slope, and plateau of Mengzi in 2018. The results show that the water required by most crops in their growth and development stage in the study area mainly depends on artificial irrigation, such as the rice, wheat, peanut, rapeseed, soybean, potato, and grape grown in Dawazi, and the apple and flue-cured tobacco grown in Duogu and Niuerpo, which does not match the local precipitation law well; only the corn and marigold grown in Duogu and Niuerpo have the ecological water requirements quota similar or nearly the same to the effective precipitation. The results imply that the agricultural structure and agricultural planting patterns need to be adjusted in this region. In conclusion, this study suggests that some crops with less water consumption and higher heat requirements are suitable for planting in the basin, while some crops with better ecological protection and economic benefits are suitable for planting in the karst mountainous areas (Zeng et al., 2020).

从合理调控生态用水角度出发，探讨典型岩溶断陷盆地区农作物生态需水问题。根据蒙自断陷盆地的岩溶地貌特征，选取位于盆地、坡面和高原面的大洼子、朵古、牛耳坡3个观测点，利用Penman-Monteith公式、作物系数（FAO推荐）及同期有效降水量，估算2018年3个观测点的参考蒸散量、生态需水量及不同作物生长所需的人工灌溉水量。结果表明，研究区大部分农作物所需的水分主要依靠人工灌溉，如种植在大洼子的水稻、小麦、花生、油菜、大豆、马铃薯、葡萄和种植在朵古及牛耳坡的苹果、烤烟，与本地降水分布规律匹配度不高，仅种植在朵古及牛耳坡的玉米和万寿菊的生态需水定额与同期有效降水之间的差值较小甚至完全满足，说明区内农业结构与种植模式有待调整。综上所述，在岩溶断陷盆地内要种植耗水较少、对热量要求较高的作物，岩溶山区则需要发展具有生态保护和经济效益的作物（曾锐等，2020）。

Comparison of water requirements and effective precipitation during each stage of crop growth

各农作物生长阶段需水量及同期有效降水量对比

6) Impact of rocky desertification control on soil bacterial community in karst faulted basin, Southwestern China
中国西南岩溶断陷盆地石漠化治理对土壤微生物群落的影响

Microorganisms play critical roles in belowground ecosystems, and karst rocky desertification (KRD) control affects edaphic properties and vegetation coverage. However, the relationship between KRD control and soil bacterial communities remains unclear. 16S rRNA gene next-generation sequencing was used to investigate soil bacterial community structure, composition, diversity, and co-occurrence network from five ecological types in KRD control area. Moreover, soil physical-chemical properties and soil stoichiometry characteristics of carbon, nitrogen, and phosphorus were analyzed. Soil N and P co-limitation decreased in the contribution of the promotion of KRD control on edaphic properties. Though soil bacterial communities appeared strongly associated with soil pH, soil calcium, soil phosphorus, and plant richness, the key factor to determine their compositions was the latter via changed edaphic properties. The co-occurrence network analysis indicated that soil bacterial network complexity in the natural ecosystem was higher than that in the additional management ecosystem. *Candidatus Udaeobacter*, Chthoniobacterales, and Pedosphaeraceae were recognized as the key taxa maintaining karst soil ecosystems in KRD control area. Our results indicate that natural recovery is the suitable way for restoration and rehabilitation of degraded ecosystems, and thus contribute to the ongoing endeavor to appraise the interactions among soil-plant ecological networks (Li et al., 2021).

微生物在地下生态系统中发挥着关键作用，并且岩溶石漠化治理能够影响土壤理化性质和植被覆盖。然而，岩溶石漠化治理与土壤微生物群落之间的关系尚不清楚。利用16S rRNA 基因下一代测序技术，在岩溶石漠化治理区对五种生态类型的土壤微生物群落结构、组成、多样性和共生网络进行了研究。此外，还分析了土壤物理化学性质和碳、氮、磷的化学计量特征。土壤氮、磷共同限制降低了岩溶石漠化治理的成效。尽管土壤微生物群落与土壤 pH、土壤钙、土壤磷和植物丰富度密切相关，但决定其组成的关键因素是土壤特性。共生网络分析表明，原生生态系统的土壤微生物网络复杂度高于次生生态系统。

Candidatus Udaeobacter、Chthoniobacterales 和 Pedosphaeraceae 是岩溶石漠化治理区岩溶土壤生态系统的关键类群。研究结果表明，自然恢复是恢复和修复退化生态系统的合适方式，也有助于评估土壤－植物生态网络之间的相互作用（Li et al., 2021）。

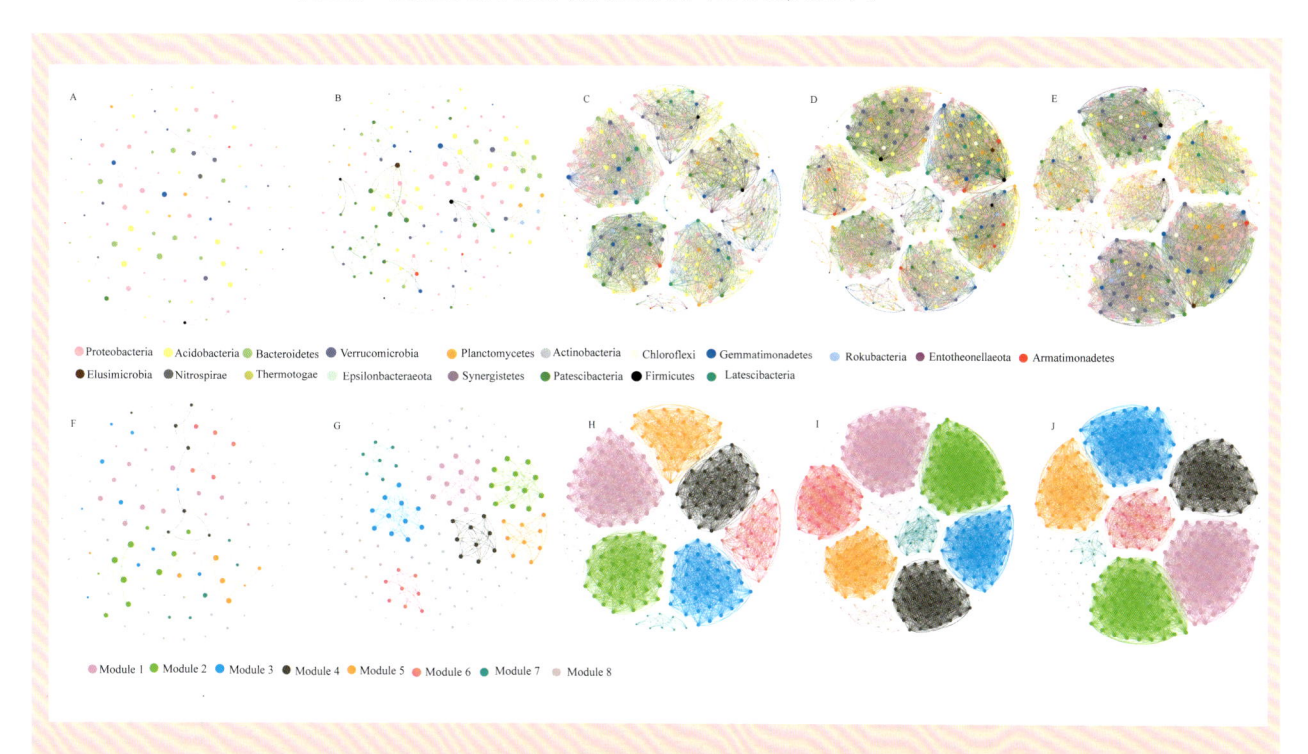

Soil bacterial co-occurrence networks in CL(A, F), GL(B, G), PF(C, H), AF(D, I), and RD (E, J), based on correlation analysis. The nodes in the network (A–E) are colored according to phylum, while the nodes in the network (F–J) are colored with respect to modularity class. The size of each node is proportional to the number of connections. (CL: corn land; GL: grassland; AF: *Alnus ferdinandi-coburgii* Schneid forest; PF: *Pinus yunnanensis* forest; RD: rock covered soil without disturbance)

基于相关性分析，图 A、F 表示 CL，图 B、G 表示 GL，图 C、H 表示 PF，图 D、I 表示 AF 和图 E、J 表示 RD 中的土壤细菌共生网络。共生网络节点（A~E）中的颜色表示门水平，而共生网络（F~J）中的颜色表征节模块类别。每个节点的大小与连接数成正比。（CL 为玉米地；GL 为草地；AF 为川滇桤木林；PF 为云南松林；RD 为未受干扰土壤覆盖区）

3. Redistribution of precipitation by vegetation and its ecohydrological effects in a typical epikarst spring catchment
典型表层岩溶泉域植被对降雨的再分配研究

Based on the analysis of the characteristics of two typical plants, the study monitored the features of the through rainfall and the stem flow of *Toona sinensis* (S1) vegetation and *Caesalpinia decapetala* (S2) vegetation, and the boreholes and epikarst water changes at the S31 karst spring in Yaji Experimental Site, Guilin. The results show that the through rainfall in S1 vegetation is 1,861.83 mm, accounting for 59.65% of the total precipitation. Through rainfall in S2 vegetation is 1,626.42 mm, accounting for 52.11% of the total precipitation. The through rainfall rate decreases with increasing precipitation. Stem flow in S1 is 89.4 mm, accounting for 2.86% of total precipitation. Stem flow in S2 is 27.79 mm, accounting for 0.89% of total precipitation. Interception storage in S1 and S2 are 1,169.97 mm and 1,466.99 mm, accounting for 37.48% and 47.01% of total rainfall respectively. Using the water balance method, the annual evapotranspiration in the Yaji typical epikarst spring catchment is 1,623.81 mm, accounting for 52.03% of the precipitation, and the runoff depth is 1,497.39 mm, accounting for 47.97% of the precipitation. The canopy could change the precipitation recharge pattern and quantity to the epikarst zone. The precipitation is intercepted by the canopy and partitioned into through rainfall and stem flow, while the through rainfall recharges the epikarst zone in a continuous wave-like manner, while the stem flow recharges in a fast-and-concentrated manner (Deng et al., 2018).

本文在表层岩溶泉域植被结构特征分析的基础上，监测桂林丫吉试验场 S31 号泉域内香椿（*Toona sinensis*）和云实（*Caesalpinia decapetala*）两种主要植被的穿透雨量和树干径流特征，以及钻孔和表层岩溶水的变化。结果表明：香椿林的总穿透雨量为 1861.83 mm，占总降雨量的 59.65%；云实灌丛总穿透雨量为 1626.42 mm，占总降雨量的 52.11%；穿透雨率随降雨量增加而减少。香椿林的树干径流总量为 89.4 mm，占总降雨量的 2.86%；云实灌丛的树干径流总量为 27.79 mm，占总降雨量的 0.89%；香椿林和云实灌丛的林冠截留总量分别为 1169.97 mm 和 1466.99 mm，平均截留率为 37.48% 和 47.01%；用水量平衡法计算得出以灌丛覆盖为主的 S31 号表层岩溶泉域年蒸散量为 1623.81 mm，占降水量的 52.03%，年径流深度为 1497.39 mm，占降水量的 47.97%。植被冠层改变了降雨对表层岩溶带的补给形式和补给量。降雨经过植被冠层的截留后转化成穿透雨和树干径流进入表层岩溶带，穿透雨以连续波状的形式补给表层岩溶带，而树干径流则以快速集中的方式补给表层岩溶带（邓艳等，2018）。

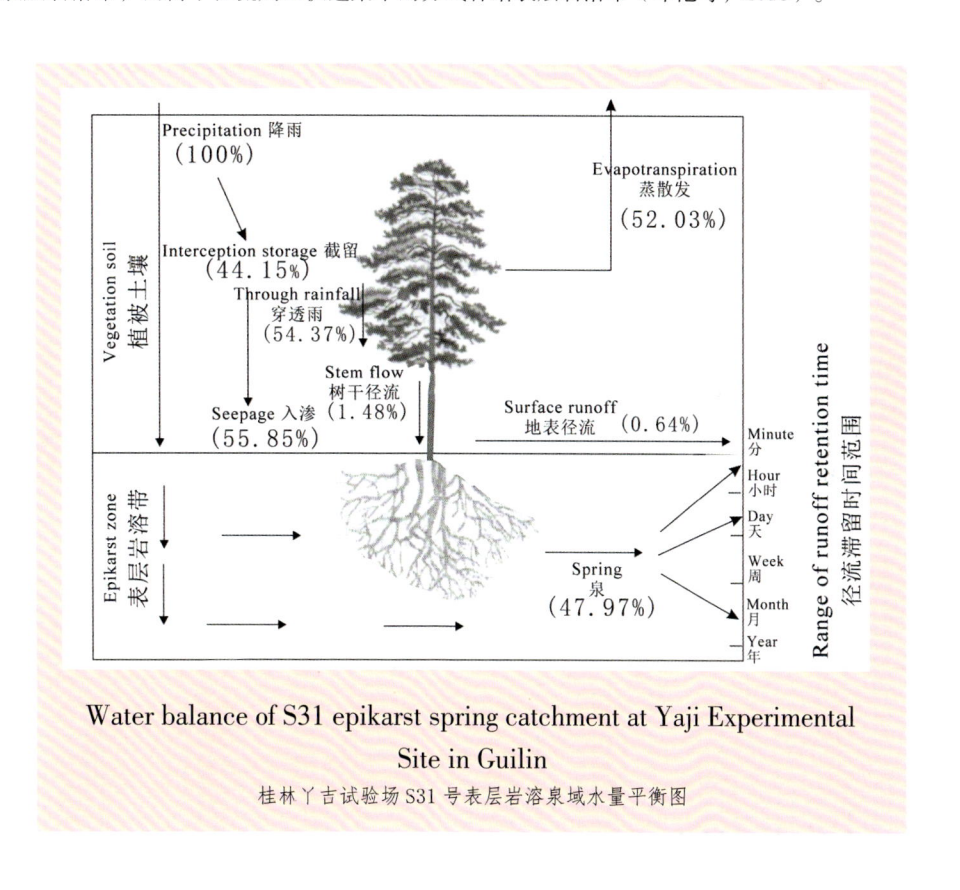

Water balance of S31 epikarst spring catchment at Yaji Experimental Site in Guilin

桂林丫吉试验场 S31 号表层岩溶泉域水量平衡图

4. The features of organic carbon of soil in karst and non-karst areas in Slovenia

斯洛文尼亚岩溶区、非岩溶区土壤有机碳组分特征

By comparing the characteristics of four soil organic carbon components (e.g., labile organic carbon, LOC; recalcitrant organic carbon, ROC; calcium bound organic carbon, Ca-SOC; iron/aluminum bound organic carbon, Fe/Al-SOC) in karst area and flysch area under the same climatic conditions in Slovenia, the results show that: ① The concentration of SOC and SOC fractions decreases with the increase of depth, indicating that the soil profiles have been stable. ② SOC values (9.7–45.5 g/kg) were consistent with the findings of other studies on soils in the region. ROC and Fe/Al-SOC (51.5%–65.8% and 68.0%–73.3%, respectively) were the major SOC fractions, while Ca-SOC accounted for a considerably lower proportion (6.4%–7.4%) of the SOC contents. ③ The key factors influencing SOC contents were calcite (expressed as calcium oxide) and clay contents, which represent mineral complexes stabilizing SOC. The overall Fe_2O_3 and Al_2O_3 concentrations did not explain differences in SOC nor its fractions, potentially due to the importance of the chemical/mineral forms of Fe-and Al-related minerals (reactivity). ④ Soils on carbonate rocks, which are richer in clay and CaO, had 6.35 g/kg (28%) higher concentrations of SOC average when compared with soils on siliciclastic rocks, due to higher concentrations of stabilized SOC fractions. The results demonstrate that bedrock lithology and pedogenesis are key factors influencing SOM stabilization (Yang et al., 2019a).

对比研究斯洛文尼亚同一气候条件下岩溶区和复理石区四种土壤有机碳(SOC)组分（分别为易氧化有机碳 LOC、难降解有机碳 ROC、钙结合态有机碳 Ca-SOC、铁/铝结合态有机碳 Fe/Al-SOC）特征，结果表明：①土壤有机碳和土壤有机碳组分浓度随深度的增加而降低，表明土壤剖面已稳定。②土壤有机碳含量（9.7~45.5 g/kg）与前人研究结果一致，其中 ROC 和 Fe/Al-SOC（分别为 51.5%~65.8% 和 68.0%~73.3%）是主要的土壤有机碳组

分，而 Ca-SOC 占 SOC 含量的比例（6.4%~7.4%）相当低。③影响土壤有机碳含量的关键因素是方解石（以氧化钙表示）和黏土含量，这代表稳定土壤有机碳的矿物复合体。总体 Fe_2O_3 和 Al_2O_3 浓度无法解释 SOC 及其组分的差异，这可能是由于 Fe 和 Al 相关矿物的化学／矿物形式（反应性）的重要性。④富含黏土和 CaO 的碳酸盐岩的土壤与硅质碎屑岩土壤相比，其土壤有机碳含量平均高出 6.35 g/kg（28%），因为稳定土壤有机碳浓度更高。结果表明，基岩岩性和成土作用是影响土壤有机碳稳定性的关键因素（Yang et al., 2019a）。

Locations of Slovenian research area and sampling sites
斯洛文尼亚研究区及采样点分布图

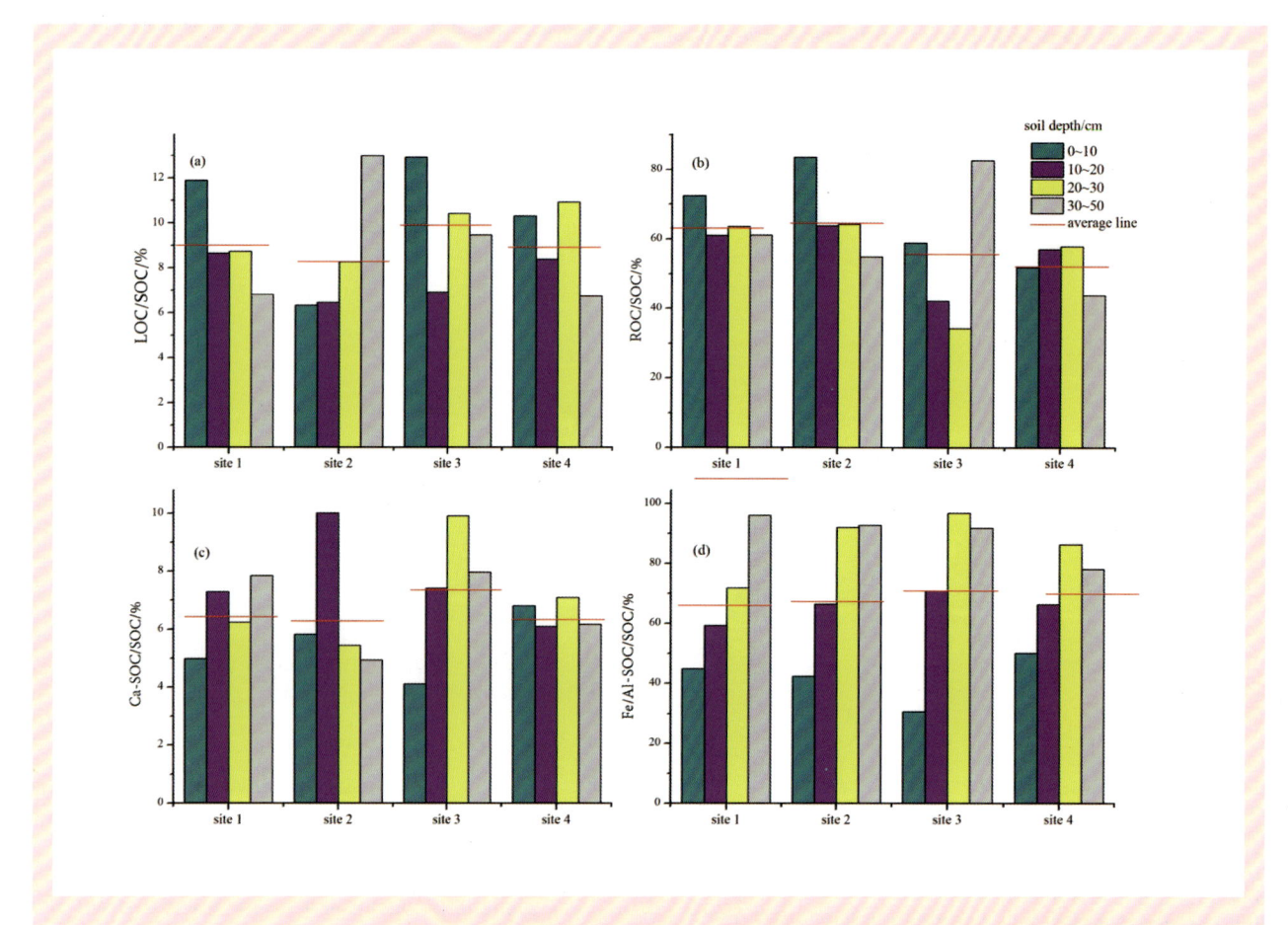

Proportions of SOC fractions in soil profiles obtained from carbonate and siliciclastic bedrocks: (a) LOC/SOC, (b) ROC/SOC, (c) Ca-SOC/SOC, and (d) Fe/Al-SOC/SOC.

碳酸盐岩基岩与硅质碎屑岩基岩土壤剖面中土壤有机碳组分所占比例

（a）易氧化有机碳／土壤有机碳；（b）难降解有机碳／土壤有机碳；（c）钙结合态有机碳／土壤有机碳；（d）铁／铝结合态有机碳／土壤有机碳

5. Decreased inorganic N supply capacity and turnover in calcareous soil under degraded rubber plantation in the tropical karst region
热带岩溶区退化橡胶林下石灰土无机氮供应能力和周转速率下降

Investigating soil inorganic nitrogen (N) supply and availability can guide rubber (*Hevea brasiliensis*) cultivation in tropical regions, but the mechanisms controlling the inorganic N supply remain unknown. In this study, three natural forests and three degraded rubber plantations located in a tropical karst region of southwestern China were sampled to determine the gross N transformation rates using a ^{15}N tracing method. Natural forests were characterized by a high soil inorganic N supply capacity and a high-level nitrate (NO_3^-) production potential, due to the high rates of organic N mineralization to ammonium (NH_4^+)(M_{Norg}) and NH_4^+ oxidation to NO_3^- (O_{NH_4}) but relatively low rates of immobilization of NH_4^+ (I_{NH_4}) and NO_3^- (I_{NO_3}) to organic N and dissimilatory NO_3^- reduction to NH_4^+ (DNRA). In the soils of the degraded rubber plantations, the rates of M_{Norg}, O_{NH_4}, I_{NO_3}, and DNRA were lower but the rates of NH_4^+ adsorption on cation-exchange sites (A_{NH_4}) increased, resulting in reductions in the inorganic N supply capacity and N availability. In addition, NO_3^- turnover in the soils of the degraded rubber plantations decreased, accompanied by a high mean residence time of NO_3^- and low δ^{15}N values. Soil total N, organic C, phosphorus, and potassium concentrations, water-holding capacity, cation-exchange capacity, and sand content were significantly lower in the soils of the degraded rubber plantations than in those of the natural forests, indicating a decline in soil quality in the former. The significant, positive relationships between these soil properties and the rates of M_{Norg}, O_{NH_4}, I_{NO_3}, and DNRA highlight the importance of the appropriate application of organic N fertilizers as well as phosphorus and potassium fertilizers to stimulate soil N cycling and thereby increase the inorganic N supply. A reduction of the N deficiency in soils used for rubber tree cultivation would alleviate the soil degradation that characterizes many rubber plantations in tropical karst regions (Garousi et al., 2021).

研究土壤无机氮的供应和有效性可以指导热带地区橡胶树（*Hevea brasiliensis*）种植，但控制土壤无机氮供应的机制尚不清楚。本研究采用 ^{15}N 标记方法研究了中国西南热带岩溶区 3 个天然林和 3 个退化橡胶林土壤氮转化过程速率。天然林土壤具有高无机氮供应能力和高硝酸盐产生潜势。这是由于有机氮矿化为铵态氮和铵态氮氧化为硝态氮速率较高,但铵态氮和硝态氮微生物同化速率和硝态氮异化还原成铵速率相对较低。退化橡胶林土壤有机氮矿化、铵态氮氧化、硝态氮微生物同化和异化还原成铵等过程速率较低,但铵态氮吸附速率增加,导致土壤无机氮供应能力和氮有效性降低。此外,退化橡胶林土壤硝态氮周转速率下降而停留时间较高,δ^{15}N 值偏低。退化橡胶林土壤总氮、有机碳、磷和钾浓度,持水能力,阳离子交换能力和砂粒含量显著低于天然林,表明退化橡胶林土壤质量下降。这些土壤性质与有机氮矿化为铵态氮、铵态氮氧化、铵态氮微生物同化和硝态氮异化还原速率呈显著正相关,表明适当施用有机氮肥及磷钾肥可以促进土壤氮循环过程,从而增加无机氮供应。降低橡胶树种植土壤中的氮缺乏将缓解热带岩溶区众多橡胶树种植园土壤退化问题(Garousi et al., 2021)。

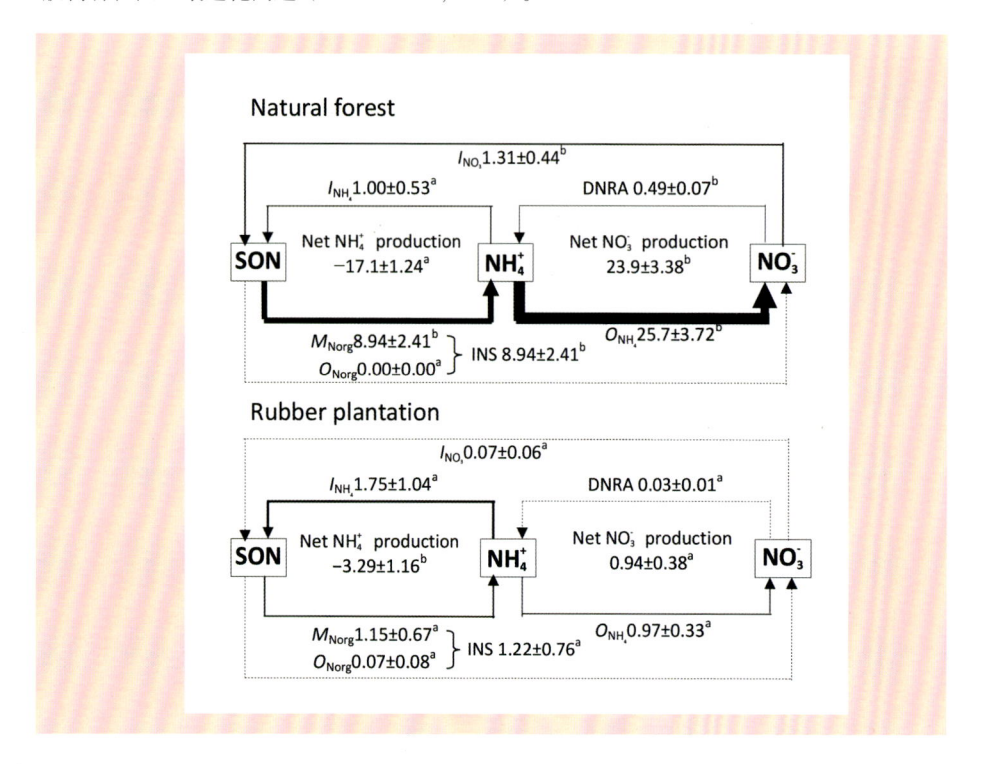

对页图 / Oppsite page

Soil N transformation rates, net NH_4^+ and NO_3^- production rates, and inorganic N supply capacity (INS)(mg N/(kg·d)) in natural forest and degraded rubber plantation soils. The arrow size is proportional to the intensity of the influence of the indicated N transformation process on N cycling. Different letters for identical N transformation rates indicate significant differences ($P=0.05$) between natural forest and rubber plantation soils. M_{Norg}, organic N mineralization to NH_4^+; I_{NH_4}, NH_4^+ immobilization to organic N; O_{Norg}, organic N oxidation to NO_3^-; O_{NH_4}, NH_4^+ oxidation to NO_3^-; DNRA, dissimilatory NO_3^- reduction to NH_4^+; I_{NO_3}, NO_3^- immobilization to recalcitrant organic N; SON, soil organic N.

天然林和退化橡胶林土壤氮转化率、铵态氮和硝态氮净产生速率及无机氮供应能力（INS）（mg N/(kg·d)）。箭头大小与指示的氮转化过程对氮循环的影响强度成正比。同一氮转化率的不同字母表明天然林和橡胶种植土壤之间存在显著差异（$P=0.05$）。M_{Norg}，有机氮矿化为铵态氮（NH_4^+）；I_{NH_4}，铵态氮固化为有机氮；O_{Norg}，有机氮氧化为硝态氮；O_{NH_4}，铵态氮氧化为硝态氮；DNRA，硝态氮异化还原成铵；I_{NO_3}，硝态氮微生物同化为有机氮；SON，土壤有机氮。

6. Bacterivore nematodes stimulate soil gross N transformation rates depending on their species

食细菌线虫种类影响土壤总氮转化率

The study conducted a microcosm experiment with soil being sterilized, reinoculated with native microbial community and subsequently manipulated the bacterivorous nematodes, including three treatments: without (CK) or with introducing one species of the two bacterivores characterized with different body sizes but similar c-p (colonizer-persister) value (*Rhabditis intermedia* and *Protorhabditis oxyuroides*, accounted for 6% and 59% of bacterivores in initially undisturbed soil, respectively). The study monitored the N_2O and CO_2 emissions, soil properties, and especially quantified gross N transformation rates using ^{15}N tracing technique after the 50 days incubation. No significant differences were observed in soil NH_4^+ and NO_3^- concentrations between the CK and two bacterivores, but this was not the case for gross N transformation rates. In comparison to CK, *R. intermedia* did not affect soil N transformation rates, while *P. oxyuroides* significantly increased the rates of mineralization of organic N to NH_4^+, oxidation of NH_4^+ to NO_3^-, immobilization of NO_3^- to organic N and dissimilatory NO_3^- reduction to NH_4^+. Furthermore, the mean residence time of NH_4^+ and NO_3^- pool was greatly lowered by *P. oxyuroides*, suggesting it stimulated soil N turnover. Such a stimulatory effect was unrelated to the changes in abundance of bacteria and ammonia-oxidizing bacteria (AOB). In contrast to CK, only *P. oxyuroides* significantly promoted soil N_2O and CO_2 emissions. Noticeably, bacterivores increased the mineralization of recalcitrant organic N but decreased soil $\delta^{13}C_{TOC}$ and $\delta^{15}N_{TN}$ values, in particular for *P. oxyuroides*. Combining trait-based approach and isotope-based analysis showed high potential in moving forward to a mechanistic understanding of bacterivore-mediated N cycling (Zhu et al., 2018).

本研究开展了土壤微宇宙试验，采用对土壤进行灭菌、重新接种土壤微生物和不同种类食细菌线虫的方法，设置三种处理：不接种食细菌线虫（CK）、接种具有不同体型但具有相似c-p（colonizer-persister）值的两种食细菌线虫（*Rhabditis intermedia* 和 *Protorhabditis oxyuroides*，分别占初始原状土壤食细菌线虫的6%和59%）。培养50天后，使用 ^{15}N 同位素标记方法测定了 N_2O 和 CO_2 的排放、土壤性质，特别是量化了土壤氮转化过程速率。不接种和接种食细菌线虫对土壤铵态氮和硝态氮浓度没有显著影响，但氮转化过程速率有较大差异。与CK相比，*R. intermedia* 不影响土壤氮转化过程速率，*P. oxyuroides* 显著提高了有机氮矿化为铵态氮、铵态氮氧化为硝态氮、硝态氮微生物同化和异化还原成铵等过程速率。此外，*P. oxyuroides* 大大降低了铵态氮和硝态氮的平均停留时间，表明这一类食细菌线虫加快了土壤无机氮周转。这种刺激作用与细菌和氨氧化细菌丰度的变化无关。与CK相比，只有*P. oxyuroides* 显著促进了土壤 N_2O 和 CO_2 的排放。值得注意的是，食细菌线虫增加了难降解有机氮的矿化，但降低了土壤 $\delta^{13}C_{TOC}$ 和 $\delta^{15}N_{TN}$ 的值，特别是*P. oxyuroides*。该研究采用特征分析与同位素分析相结合的方法有助于深入了解食细菌线虫影响下的氮循环机制 (Zhu et al., 2018)。

Effect of bacterivores on soil N transformation rates (mg N/(kg·d)) after 50-d incubation (n=4, ±SD). The arrow width represents the size of the corresponding gross N transformation rate. For the same N transformation rate in different treatments, different letters indicate significant differences among the three treatments at P = 0.05. M_{Norg} is the mineralization of organic N to NH_4^+; I_{NH_4} is the immobilization of NH_4^+ to organic N; O_{NH_4} is the oxidation of NH_4^+ to NO_3^-; O_{Nrec} is the oxidation of recalcitrant organic N to NO_3^-; I_{NO_3} is the immobilization of NO_3^- to recalcitrant organic N; DNRA is dissimilatory NO_3^- reduction to NH_4^+. Different small letters for the same N transformation rate indicated significant difference among different treatments at 0.05 level.

培养50天后食细菌线虫对土壤氮转化过程速率的影响（mg N/(kg·d)，n=4，±SD）。箭头宽度表示相应的氮转化过程速率大小。对于不同处理土壤中相同的氮转化过程速率，不同字母表示三种处理之间显著差异（P<0.05）。M_{Norg}，有机氮矿化为铵态氮；I_{NH_4}，铵态氮微生物同化为有机氮；O_{NH_4}，铵态氮氧化为硝态氮；O_{Nrec}，难分解有机氮氧化为硝态氮；I_{NO_3}，硝态氮微生物同化为有机氮；DNRA，硝态氮异化还原成铵。

7. National Field Scientific Observation and Research Station of Karst Ecosystem in Pingguo, Guangxi
广西平果喀斯特生态系统国家野外科学观测研究站

In 2020, the Field Scientific Observation and Research Station of Karst Ecosystem in Pingguo, Guangxi was approved by the Ministry of Science and Technology to be included in the list for preferential construction of national field stations. It is also the core station for ecological environment monitoring in karst areas. The station formed a network by covering typical karst types in southwest China with the karst rocky desertification field observation station in Pingguo City, Baise City, Guangxi as the main station, and the stations in Guilin, Guangzhou, Wulong, Mengzi, and Libo as substations. The station carries on long-term continuous observation, experiments, research, and demonstration about karst rocky desertification and its resources and environmental effects. Recently, the station, orienting towards the frontier of national major strategies and disciplinary development, has focused on: ① Basic research on karst rocky desertification observation: typical station construction – screening of monitoring indicator systems – development of modern monitoring technologies – monitoring of material and energy transfer and transformation processes and evaluation of resources and environmental effects. ② The formation of typical karst environmental problems such as karst rocky desertification: the formation mechanisms and resources environmental effects of rocky desertification formed in different types of environment; the formation mechanism and process of problems such as water and soil erosion and leakage, karst collapse, and landscape degradation. ③ Comprehensive treatment of typical karst environmental problems such as karst rocky desertification: demonstration of comprehensive control of different environmental types of rocky desertification; monitoring of soil and water leakage and demonstration of soil and water conservation experiments; karst groundwater dynamic monitoring and demonstration of experiments on drought/flood regulation and control; monitoring of carbon-water-calcium cycle and demonstration of experiments on carbon sink increase.

广西平果喀斯特生态系统国家野外科学观测研究站于 2020 年经科技部批准列入国家野外站择优建设名单，是岩溶区生态环境监测的核心台站。该站以建在广西百色市平果市的岩溶石漠化野外观测站为主站，并与设在桂林、广州、武隆、蒙自、荔波的 5 个子站，形成覆盖西南典型岩溶类型区的观测研究站网，围绕岩溶石漠化及其资源环境效应，开展长期稳定连续的观测、试验研究和科技示范。近些年来，该站面向国家重大战略和学科发展前沿，着重开展：①岩溶石漠化观测基础研究：典型站建设 – 监测指标体系筛选 – 现代化监测技术研发 – 物质能量迁移转化过程监测及其资源环境效应评价。②岩溶石漠化等典型岩溶环境问题形成：不同环境类型石漠化形成机理及其资源环境效应；水土漏失、岩溶塌陷、景观退化等问题的形成机理和过程。③岩溶石漠化等典型岩溶环境问题综合治理：不同环境类型石漠化综合治理示范；水土漏失监测及水土保持试验示范；岩溶地下水动态监测及旱涝调控试验示范；碳 – 水 – 钙循环监测及增汇试验示范。

The station has over 200 sets of instruments and equipment, including a critical zone monitoring integration system, groundwater automatic monitoring instrument, water and soil non-point source pollution investigation and monitoring integration system, multi-channel soil carbon flux automatic monitoring system, multi-channel karst collapse dynamic monitoring system, etc. It has 208 observation sites for epikarst springs, water loss and soil erosion, underground rivers, overland flow, natural water sites, etc. The monitoring contents include rocky desertification dynamics, karst hydrochemistry, water loss and soil erosion, collapse, soil carbon flux, etc. The data are reliable, which can meet the requirements of the observation indicator system in this field. There are 23 thematic experimental sites, including water and soil leakage prevention and control experimental sites, slope water and soil leakage prevention and control experimental sites, soil improvement experimental sites, groundwater tracing experimental stations, epikarst water circulation and regulation function experimental sites, and karst fissure water and conduit water exchange experimental sites, which can meet the needs of karst environmental experiments and research (Li et al., 2021).

该站拥有关键带监测集成系统、地下水自动监测仪、水土面源污染调查监测集成系统、多通道土壤碳通量自动监测系统、多通道岩溶塌陷动力监测系统等共计 200 多台（套）仪器设备，建有表层岩溶泉、水土流失、地下河、坡面流、天然水点等208个观测点，监测对象包括石漠化动态、岩溶水化学、水土流失、塌陷、土壤碳通量等，数据精度具有可靠保障，可满足本领域观测指标体系要求。建有水土漏失防治试验区、坡面水土漏失防治试验区、土壤改良试验区、地下水示踪试验站、表层岩溶带水循环及调蓄功能试验场、岩溶裂隙与管道水流交换试验场等专题试验场23个，能够满足岩溶环境试验研究的需求（李文莉等，2021）。

Experimental areas of the National Field Scientific Observation and Research Station of Karst Ecosystem in Pingguo, Guangxi
广西平果喀斯特生态系统国家野外科学观测研究站试验区

Base of the National Field Scientific Observation and Research Station of Karst Ecosystem in Pingguo, Guangxi
广西平果喀斯特生态系统国家野外科学观测研究站基地

3.2.4 Karst geoheritage diversity and sustainable development
岩溶地质遗迹多样性和可持续开发

1. Assisted the successful application of the first UNESCO Global Geopark (UGGp) in Thailand—Satun UGGp
   助力泰国成功申报首个世界地质公园——沙敦世界地质公园

In May 2017, a scientist delegation led by Mr. Chen Weihai from IRCK went to Thailand to provide technical services for the application of Satun Geopark in Thailand to be listed as an UGGp. IRCK delegation verified and evaluated the application text, signs, popular science interpretation, community participation, popular science education and management agencies of Satun Geopark according to the criteria of UNESCO. They proposed that the application should focus on the demonstration and international comparison of the world-class geology and its values of Satun Geopark, highlighting the rules and representativeness of the karst development on continents, along coastal lines and in the ocean, as well as the uniqueness of the development of surface and subsurface karst in tropical areas, so that the whole values of Satun Geopark could be presented well. The suggestions were highly recognized by Thai side, who revised their application accordingly. At last, they passed the evaluation of UNESCO and listed as the first UNESCO Global Geopark in Thailand.

2017年5月，中心科技人员陈伟海等一行三人赴泰国为沙敦世界地质公园申报提供技术服务，我方根据联合国教科文组织的标准对泰国沙敦世界地质公园申报文本、标示标牌、科普解说、社区参与、科普教育和管理机构进行了核实和评价，提出了泰国沙敦地质公园申报应主要解决其世界级地质价值的论证和国际对比，抓住泰国沙敦地质公园陆地、海岸带和海上岩溶系统发育的规律和代表性，以及热带地区地表、地下岩溶发育的独特性，从而深挖地质公园的整体价值。我方建议受到了泰方的高度认可，泰方经过后续调整，成功通过了联合国教科文组织的专家评审，成为联合国教科文组织世界地质公园。

Chapter 3 Scientific Research

The in-door discussion for then Satun aspiring UGGp application
沙敦地质公园迎检前室内研讨

The field investigation of IRCK delegation in then Satun aspiring UGGp
中心团队在沙敦地质公园开展野外调查

2. Carried out continuous research after successful support to the application of Zhijindong Cave UGGp: the karst landscape formation mechanism and model of the Zhijindong Cave Global Geopark, Guizhou Province

开展贵州织金洞世界地质公园申报成功后的科学研究：贵州织金洞世界地质公园岩溶成景机制及模式研究

Zhijindong Cave UNESCO Global Geopark, located in the west of Guizhou Plateau, is located in Zhijin County and Qianxi City, Bijie City, Guizhou Province. It is divided into Zhijindong Cave Park (15.7 km^2), Qijiehe River Park (45.8 km^2), and Dongfenghu Lake Park (108.5 km^2), which are consistent with three geomorphic units representing different karst development processes. In 2015, IRCK supported Zhijindong Cave Geopark to apply for global geopark successfully, and continued to carry out related research on Zhijindong Cave UGGp. The related research results showed that Zhijin karst geoheritage could be classified into eight categories: cave, gorge, natural bridge, tiankeng, high fengcong, hill, cuesta and pictographic mountain, hydrological heritage. All of them have been developed in the Lower Triassic marine carbonate strata and distributed orderly and intensively in such three karst geomorphic units as Zhijindong Cave, Qijiehe River, and Dongfenghu Lake, which are relatively separate and complete but closely linked to each other through Qijiehe River. These karst landscapes constitute a majestic, typical, beautiful, and precious plateau karst landscape group with caves, gorges, natural bridges, and tiankengs (giant sinkholes) as the most important highlights. Based on the analysis and investigation about the regional background of the landscape formation, it is indicated that the evolution of Zhijin karst beginning in Paleogene with the hydraulic connection linking the three karst areas as the major axis, went through four stages: initial stage (karst outcrop in Paleogene)→embryonic stage (split-axis discrete island-style landscape formation process)→significant development stage (principal-axis discrete island-style and collective island-style landscape formation process)→modern karst stage. Based on the above analysis, four island-style landscape formation mechanisms are put forward, namely, split-axis type, principal-axis type, discrete type, and collective type. At the same time, the Zhijin karst landscape formation modes are generalized into the "adjoining island-style landscape formation mode" after comparing with other similar karst areas (such as Leye in Guangxi, Wulong in Chongqing, etc.)(Wei et al., 2018).

织金洞世界地质公园，地处贵州高原西部，位于贵州省毕节市织金县和黔西市境内，划分为织金洞（15.7 km²）、绮结河（45.8 km²）和东风湖（108.5 km²）三个园区，三个园区分别与三个不同岩溶发育阶段的地质单元相对应。2015年，中心支撑贵州织金洞地质公园顺利申报世界地质公园，并持续开展织金洞世界地质公园的相关研究工作。相关研究结果表明，织金岩溶地质遗迹可分为8大类：洞穴、峡谷、天生桥、天坑、高峰丛、丘陵、单面山与象形山、水文遗迹，它们以下三叠统海相碳酸盐岩为物质基础，有序、集中分布于织金洞、绮结河、东风湖三片相对独立，却又以绮结河为纽带紧密相连的岩溶区域内，共同构成一个以洞穴、峡谷、天生桥、天坑为核心，形态雄伟、典型、优美、珍稀的高原岩溶景观群。同时，在对公园区域成景背景进行分析和研究的基础上，认为织金岩溶的形成演化始于古近纪，以三处岩溶区域相互之间水力联系的演变为主轴，历经四个阶段：初始阶段（古近纪期间，岩溶地层出露）→雏形阶段（分轴型离散岛屿式成景过程）→重要发育阶段（主轴型离散岛屿式与集合型岛屿式成景过程）→现代岩溶阶段。在上述分析的基础上，提出分轴型、主轴型、离散型和集合型四种岛屿式成景机制。同时，通过与相似岩溶区域（如广西乐业、重庆武隆等）的成景模式进行对比，将织金岩溶成景模式归纳为"相邻岛屿式成景模式"（韦跃龙等，2018）。

对页图 / Oppsite page

Schematic diagrams of the formation and evolution of Zhijin Karst and Zhijindong Cave
织金喀斯特及织金洞形成演化示意图

第三章 科学研究

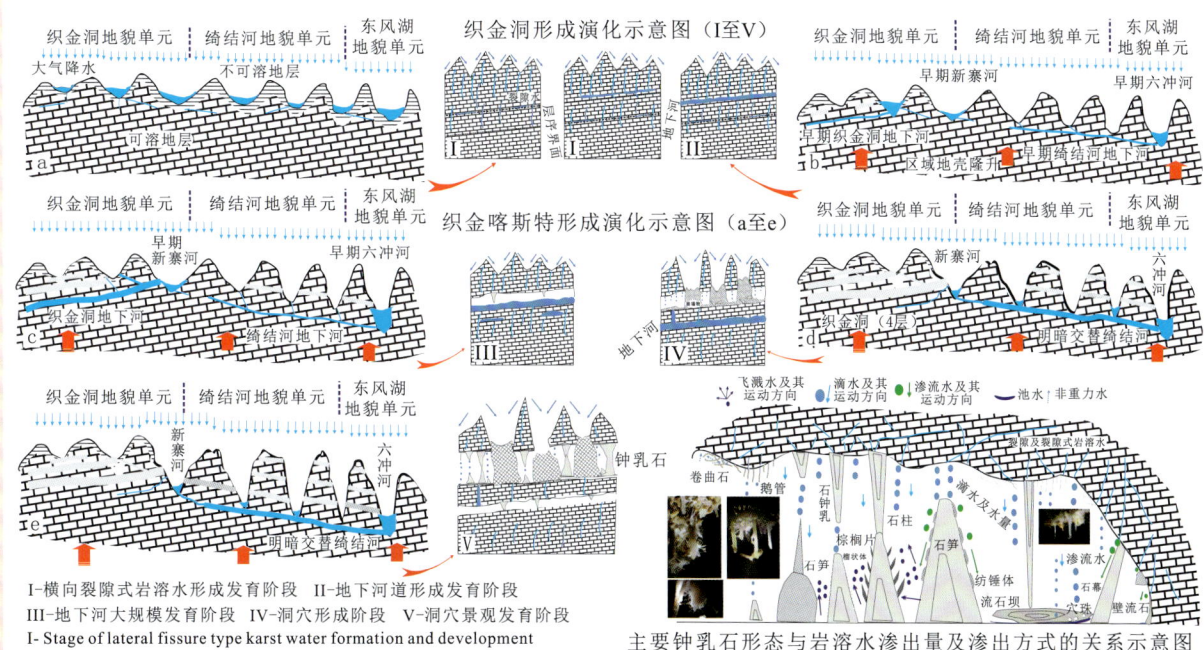

织金洞形成演化示意图（I至V）

织金喀斯特形成演化示意图（a至e）

主要钟乳石形态与岩溶水渗出量及渗出方式的关系示意图

I-横向裂隙式岩溶水形成发育阶段　II-地下河道形成发育阶段
III-地下河大规模发育阶段　IV-洞穴形成阶段　V-洞穴景观发育阶段
I- Stage of lateral fissure type karst water formation and development
II- Stage of underground river formation and development
III-Stage of underground river massive development　IV-Stage of caves formation　V-Stage of cave landscape development

a-最初大气降水沿各个低地汇聚，形成众多小型地表河；局部沿裂隙下渗，可溶岩内部进行分散、独立的侵蚀和流动。
b-受区域间歇性地壳隆升控制，众多小型地表河经不断袭夺、下潜和侵蚀，形成早期织金洞、绮结河和六冲河的主河道。
c-区域地壳隆升，各地下河和地表河继续下潜，其中织金洞地下河袭夺古新寨河，补给量陡增。
d-随后，随着新寨河被绮结河袭夺，织金洞地下河的补给量由多变少；其间，受多期区域间歇性隆升的控制，地下河通道经历多次抬升，形成织金洞的四层洞道。
e-绮结河经过多次袭夺、改道和下潜，由地下河转为明暗相间的河流，同时逐渐将织金洞、东风湖两个相对独立的单元联结起来。而六冲河等地表河为适应间歇性下降的区域侵蚀基准面，历经漫长、持续的纵向侵蚀等作用，形成河谷。

a-At the beginning, atmospheric precipitation gathered in lower place, forming many small surface rivers; some water infiltrated along fissures and dispersed, eroded and flowed inside the soluble rock independently.
b-Controlled by regional intermittent crust being recoiled, many small surface rivers formed the main riverway of Zhijindong Cave, Qijiehe River and Liuchonghe River after continuously being captured, submerging and erosion.
c-Regional crust was recoiled, all underground rivers and surface rivers continued to submerge, during this process, Zhijindong Cave underground river captured ancient Xinzhaihe River, resulting in a sudden increase of its supplementary amount.
d- Subsequently, the supplementary amount of Zhijindong Cave underground river decreased and gradually became weak because Xinzhaihe River was captured by Qijiehe River; during this process, controlled by many regional intermittent crust being recoiled, the underground river passage was raised upwards for many times, forming the 4th layered cave passage of Zhijindong Cave.
e-After being captured, diverted and submerging for many times, Qijiehe River turned to a river running underground and aboveground alternatively from underground river, meanwhile, it linked the two independent units (Zhijindong Cave and Dongfenghu Lake) together. As for Liuchonghe River, in order to suit for the intermittently decreasing erosion base, it formed a river valley after going through long and continuous vertical erosion.

国际岩溶研究中心第二个六年历程

3. Supported the successful application of Xiangxi UGGp and promoted the related research actively

支撑湘西世界地质公园成功申报，积极推进后续科学研究工作

Xiangxi UNESCO Global Geopark is located in the middle part of Hunan Province and has a total area of 2,710 km^2. The park is located in the slope zone on the eastern edge of the Yunnan-Guizhou karst plateau. It is mainly characterized by geological relics such as GSSPs—Guzhangian Stage and Paibian Stage, the world's largest red carbonate rock forest and the magnificent plateau cutting platform-canyon group. In 2020, IRCK supported Xiangxi Geopark to apply for a global geopark successfully and continues to carry out related research.

湘西世界地质公园位于湖南省中部，总面积为2710 km^2。公园地处云贵岩溶高原东部边缘斜坡地带，以全球寒武系标准层型剖面——古丈阶与排碧阶"金钉子"，世界上规模最大的红色碳酸盐岩石林和蔚为壮观的高原切割型台地－峡谷群等地质遗迹为主要特色。2020年，中心支撑湘西地质公园顺利申报世界地质公园，并持续开展湘西世界地质公园的相关研究工作。

1) Causes of formation and geo-scientific significance of karst gorge group in Xiangxi Geopark

湘西地质公园岩溶峡谷群成因及其地学意义

Xiangxi Geopark has the most densely distributed area of karst gorges in the world with more than 200 karst gorges in an area of about 1,162 km^2, which distribute like a spider network. The karst gorges appear in a manner of linear, V-shaped and U-shaped distributions, with a lot of caves, waterfalls, and stone pillars along the rock cliffs. Between gorges, there are karst platform landforms and peak cluster mountains, showing a peculiar and spectacular landscape. The alternating extension of silicate carbonate rocks and argillaceous carbonate rocks, dense tectonic features, high degree joints, inclined uplift of earth blocks and strong karst hydrodynamics together provide favorable conditions for the development of karst gorge groups. The formation of the karst gorge group can be divided into three stages: geological tectonic uplift, river gorge formation and karst gorge development. Based on the comparison with other gorges of karst areas in the world, the karst gorge group in Xiangxi is the most typical karst with the largest scale, the densest gorges, and the most typical karst in the world. Meanwhile, the karst gorge group in Xiangxi has not only important research and geological historical values, but also the great significance of geological landscape and geological cultures (Jiang et al., 2019).

湘西地质公园岩溶峡谷分布面积约 1162 km², 岩溶峡谷 200 多条, 密如蛛网分布, 是世界上最密集的岩溶峡谷群分布区。湘西岩溶峡谷形态上分为线型峡谷、"V"字形峡谷和箱形峡谷, 其间分布岩溶台地和峰丛山地, 岩溶峡谷两侧岩溶洞穴、瀑布和石柱发育, 景观奇特壮观。相间分布的硅质碳酸盐岩和泥质碳酸盐岩、密集分布的地质构造及高角度节理裂隙、地块掀斜式抬升和强岩溶水动力为岩溶峡谷群的发育提供了有利条件。岩溶峡谷群的形成经历了地质构造抬升、河流峡谷形成、岩溶峡谷发育三个阶段。与世界其他岩溶区峡谷对比, 湘西岩溶峡谷群是世界地质公园中分布规模最大、峡谷最密集、岩溶发育最典型的岩溶峡谷区。湘西岩溶峡谷群不但具有重要的岩溶学研究和地质历史价值, 而且具有重要的地质景观和地质文化意义(蒋忠诚等, 2019)。

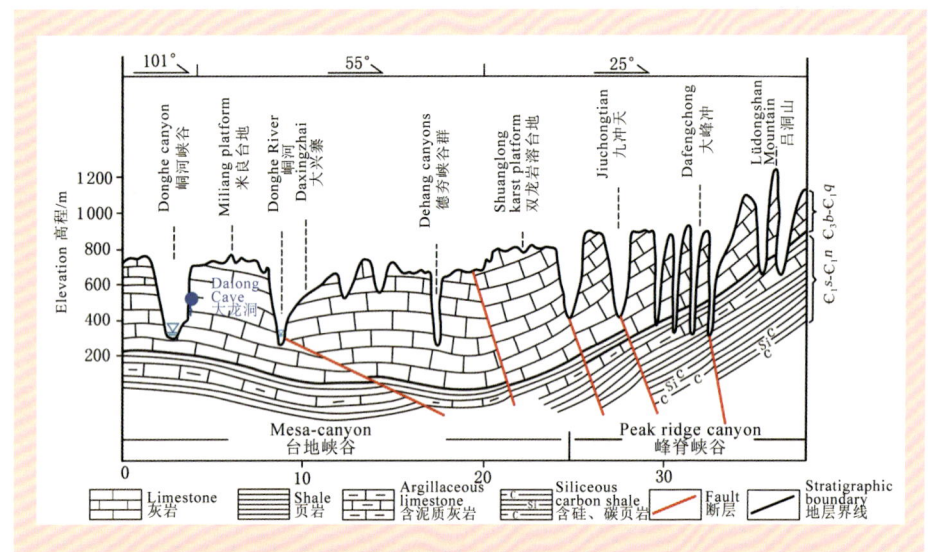

Profile of karst canyon groups geomorphology from Donghe River to Lüdongshan Mountain in Xiangxi
湘西峒河—吕洞山岩溶峡谷群地貌剖面

Karst gorge and tableland landscape in Dehang, Xiangxi
湘西德夯岩溶峡谷及台地景观

2) Geodiversity, geotourism, geoconservation, and Sustainable Development in Xiangxi UNESCO Global Geopark—A case study in ethnic minority areas

湘西世界地质公园的地质多样性、地质旅游、地质保护与可持续发展——以少数民族地区为例

Xiangxi UGGp with outstanding geoheritage resources and a long history of Tujia and Miao minorities' cultures, combines the extraordinary beauty of man and nature. In Xiangxi UGGp, there are 6 types of geosites, with 90 individual geosites including 4 of international significance, which are from the Guzhangian Stage and Paibian Stage of Global Standard Stratotype–Section and Points (GSSPs) in the Cambrian System, the world's largest karst red stone forest landscape, and a spectacular karst platform-canyon landscape. By classifying all the geoheritage sites and creating different levels of protection measures, the geopark makes a clear distinction between utilization and protection. Because of the distinctive and closed geological background, the Tujia and Miao ethnic minorities in the geopark have formed a unique lifestyle over the past thousand years. With the construction of geoparks, ethnic minorities have been able to promote their lifestyles and be lifted out of poverty through geotourism. Xiangxi UGGp maintains and supports the right of decision-making by the ethnic minority people. Therefore, the people are truly willing to take initiatives in the management of the parks. This is a successful case study of participation in geoparks by ethnic minorities. A future long-term development plan has been designed and resilient systems are formed. Xiangxi UGGp might promote ethnic minority areas better after joining UNESCO Global Geoparks and may provide a reference for other similar areas (Wu et al., 2021).

湘西世界地质公园拥有杰出的地质遗迹资源和悠久的土家族和苗族文化历史，融合了人与自然的非凡之美。湘西世界地质公园共有6种类型的地质遗迹，90个地质遗迹点，其中4个具有国际意义，即寒武系古丈阶全球标准层型剖面（Guzhangian GSSP）和寒武系排碧阶全球标准层型剖面（Paibian GSSP），世界上面积最大的岩溶红石林景观及蔚为壮观的岩溶台地-峡谷景观。公园通过对所有地质遗迹点进行分类，制定了不同级别的保护措施，明确区分了利用和保护的方式。地质公园内的土家族和苗族由于其独特而封闭的地质背景，在过去的千年里形成了独特的生活方式。随着地质公园的建设，这些少数民族得以通过地质旅游推介他们的生活方式并摆脱贫困，湘西世界地质公园也维护并支持少数民族人民的决策权,因此,当地居民愿意管理公园。这是少数民族参与地质公园的一个成功案例。湘西世界地质公园已经设计了未来的长期发展计划,并形成了弹性机制。加入联合国教科文组织世界地质公园后,其能更好地促进少数民族地区的发展,并为其他类似区域提供范式(Wu et al., 2021)。

对页图 / Oppsite page

Characteristics and geological genesis of red stone forest. (a) The uneven stacking state is formed by the difference of components. (b) There is little faulting in red stone forest area, which is mainly affected by two groups of joints. Among them, the shear joints are 30° to 45°, and the tensile joints are 315°±20°. (c) The top of the stone pillars are flake or sword shaped due to the influence of atmospheric precipitation. (d) Typical flaming stone pillars in the red stone forest. The top is sharp and the bottom is smooth. (e) Tower-shaped stone pillars. (f) The micrograph of red stone forest. Nodulars are well-developed in the rocks. (g) The formation mode of red stone forest. Paleo-sedimentary environment and tectonic uplift-hydrological conditions are the most important

红石林的特征及地质成因。(a) 不均匀的堆叠状态是由沉积岩层的组分差异造成的。(b) 红石林区断层活动较少,主要受两组节理影响。其中,剪性节理为30°~45°,张性节理为315°±20°。(c) 由于大气降水的影响,石柱顶部呈刃状或剑状。(d) 红色石林中典型的火焰似的石柱,顶部锋利,底部光滑。(e) 塔形石柱。(f) 红石林的显微照片,岩石中结核发育良好。(g) 红石林的形成模式,古沉积环境和构造抬升水文条件是最重要的。

第三章 科学研究

4. Supported the construction of Guilin Innovative Demonstration Area for the Sustainable Development Agenda
支撑桂林可持续发展议程创新示范区建设

In 2018, Guilin was approved as a National Innovation Demonstration Area for the Sustainable Development Agenda (Guilin Demonstration Area). Based on expertise advantages, IRCK actively contributed to the construction of Guilin Demonstration Area, concluded the distribution feature of natural resources in Guilin, and presented 29 maps of 6 categories to the Guilin municipal government, including natural geography and geological background, mineral resources and geological disasters, water resources, land resources, forest and grassland resources, geological heritage and tourism resources. Meanwhile, IRCK proposed the key technologies for erosion slope runoff control and aquatic ecosystem restoration in the Lijiang River Basin, established the real-time optimal flood control operation decision-making system for reservoirs in the upper reaches of the Lijiang River as well as the water conservation and navigation mode as "reservoir water recharge–karst seepage prevention–underground water storage–ecological restoration" in dry season.

2018年，桂林获批国家可持续发展议程创新示范区（桂林示范区），中心结合自身业务优势，积极投身桂林示范区建设，查明桂林市自然资源分布，向桂林市政府赠送了自然地理与地质背景、矿产资源与地质灾害、水资源、土地资源、林地和草地资源、地质遗迹和旅游资源6类29幅图；同时，提出了漓江流域侵蚀坡面径流调控和水生生态系统修复关键技术，建立了漓江上游水库群防洪实时优化调度决策系统与"水库补水－岩溶防渗－地下蓄水－生态修复"枯水期保水通航模式。

IRCK project team has carried out the research on the ecological evolution and driving mechanism of the Lijiang River Basin landscape, revealing the evolution trend and spatiotemporal evolution laws of the karst landscape pattern in the Lijiang River Basin under the impacts of climate change and human activities; it has clarified the influence of agricultural construction projects on ecological landscapes, the coupling mechanism between landscape degradation and economic development, and the comprehensive driving mechanism; and it has also proposed an optimized allocation plan and collaborative management strategy for landscape resources in river basins. The team has established three demonstration areas for small river basins (i.e. Huixian in Lingui, Xiling in Gongcheng, and Guanyan in Yanshan), as well as three demonstration sites (i.e. Yanguan in Xing'an, Lianhua in Gongcheng, and Zengpiyan in Xiangshan), covering the area of 4,312 mu (1 mu= 0.067 hectares), with the area for demonstrating the related technology as about 1,000 mu. Meanwhile, the team has cultivated six characteristic ecological industries, which has generated approximately 12 million yuan economic benefits resulted from ecological tourism annually, and increase the per capita income of residents in the demonstration areas by more than 10%. Moreover, the water ecology of the demonstration area has significantly improved, with plant diversity increased significantly, and vegetation coverage increased significantly. In addition, the team has provided 25 key technologies for the sustainable utilization of karst wetland water resources, plant landscape resources, abandoned quarry landscape restoration, and the sustainable utilization of mountain-water-farmland-forest-lake community. Importantly, the team has combined the SDGs with China's reality and evaluated the progress for sustainable development in Guilin, with an evaluation system applicable for Guilin established. The research work led by IRCK has enlightened systematic solutions for the sustainable utilization of karst landscape resources in the Lijiang River Basin and supported the construction of Guilin Innovative Demonstration Area for the Sustainable Development Agenda effectively.

中心科研团队开展漓江流域景观生态演变及退化驱动机制研究，揭示了气候变化和人类活动双重作用下的漓江流域岩溶景观格局演化态势与时空演变规律；阐明了漓江流域农业建设工程对生态景观的影响、景观退化与经济发展的耦合机理及综合驱动机制；提出了流域景观资源优化配置方案和协同管理策略。团队建设了临桂会仙、恭城西岭、雁山冠岩3个小流域示范区，以及兴安严关、恭城莲花和象山甑皮岩3个示范点，总面积4312亩，推广应用面积约1000亩；培育形成6个特色生态产业，形成每年共约1200万元的生态旅游规模效益，示范区居民人均收入增加10%以上；同时，示范区水生态明显好转，植物多样性显著增加，植被覆盖率明显提高。此外，团队为岩溶湿地水资源可持续利用、植物景观资源可持续利用、废弃采石场景观修复、"山－水－田－林－湖"景观资源共同体可持续利用提供关键技术 25 项。重要的是，开展了 SDGs 中国本土化及可持续发展进展评估，构建了桂林市本地化 SDGs 评估方法与指标体系。本研究为漓江流域岩溶景观资源可持续利用提供了系统性解决方案，有效支撑了桂林市国家可持续发展议程创新示范区建设。

Guilin—the perfect combination of karst peak clusters and rivers in South China Karst
桂林 —— 中国南方岩溶峰丛与河流的完美结合

第三章　科学研究

5. Supported the sustainable development of Wulong World Natural Heritage Site
支撑武隆世界自然遗产地可持续发展

In order to support the sustainable development of Wulong World Natural Heritage Site, IRCK has carried out serial assessment of the infrastructure constructions impact such as expressways, reservoirs, and tourism towns in Wulong to the natural reserves including the heritage sites, scenic spots, geoparks, etc. More than 30 assessment reports have been prepared, providing a scientific basis for local government to grant administrative permissions. The popular science book *World Heritage in Wulong* was compiled and published. It explained the formation process, landscape characteristics, and "outstanding significance and universal values" of Wulong Karst in plain words. It raised people's awareness to protect the world heritage. By the end of 2020, there were 4,583 rural tourism reception households in Wulong, with per capita net income of farmers engaged reached 41,000 yuan, increased by 11.4 times compared to 3,300 yuan in 2007. The annual visitors reached 37.01 million, with the comprehensive tourism income reaching 18 billion yuan, accounting for more than 90% of the region's GDP.

为支撑武隆世界自然遗产地的可持续发展，中心对武隆涉及遗产地、风景名胜区、地质公园等自然保护地的高速公路、水库、旅游小镇等基础设施项目开展了影响评估工作，编写评估报告30余份，为当地政府部门做出的行政许可提供了科学依据。编写出版了科普书籍《世界遗产在武隆》，以通俗易懂的语言阐述了武隆喀斯特世界自然遗产的形成过程、景观特征及"突出意义和普遍价值"，提醒人们要保护好世界遗产。截至2020年底，武隆乡村旅游接待户达到4583户，从事涉旅服务的农民人均纯收入达4.1万元，比2007年的0.33万元增长了11.4倍。全年旅游人数达到3701万人次，旅游综合收入为180亿元，占全区生产总值比重的90%以上。

第三章　科学研究

Wulong Jingkou Tiankeng (photoed by Mr. Cai Shi)
武隆箐口天坑（拍摄者：蔡石）

Statistics showing the growth of tourists in Wulong, Chongqing
重庆武隆区游客增长统计图

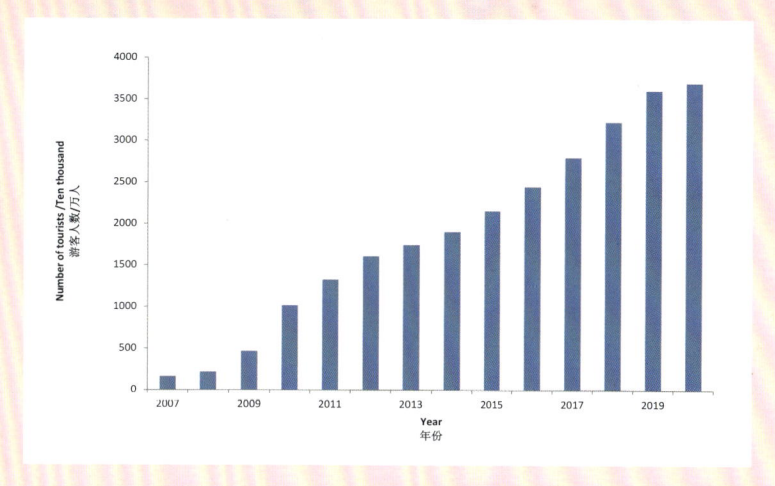

国际岩溶研究中心第二个六年历程

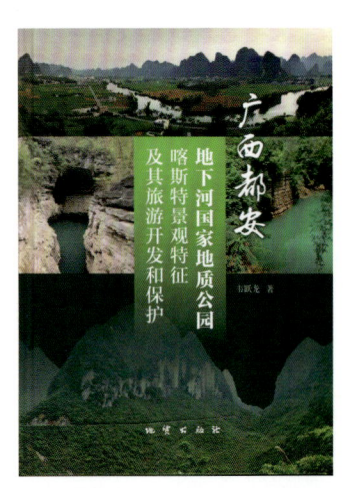

Published 7 monographs: *Karst Landform and Caves in Shandong*; *Guilin—Research on Karst Geological Resources Cultural Heritage*; *Karst Landscape Features and the Tourism Development and Protection of Du'an Underground River National Geopark, Guangxi*; *Cave Landscape and Formation Environment of Baiyun Cave, Lincheng, Hebei*; *Granite Landscape Features and the Tourism Development and Protection of Wuhuangshan National Geopark, Pubei, Guangxi*; *Guilin Lanscape*; *Leye Tiankeng*

出版了7本专著：《山东岩溶地貌与洞穴》《桂林岩溶地质资源文化遗迹研究》《广西都安地下河国家地质公园喀斯特景观特征及其旅游开发和保护》《河北临城白云洞洞穴景观及形成环境》《广西浦北五皇山国家地质公园花岗岩景观特征及其旅游开发和保护》《桂林山水》《乐业天坑》

3.2.5 Karst collapse and its prevention
岩溶塌陷与防治

1. Review of the advanced monitoring technology of groundwater-air pressure (enclosed potentiometric) for karst collapse studies
岩溶塌陷地下水 – 气压力先进监测技术综述（封闭电位法）

The study presents an overview of the groundwater-air pressure monitoring technology for karst sinkholes. Karst collapses often occurred rapidly without prior warning. The early warning for potential karst collapses has been one of the most challenging problems around the world. In fact, many karst collapses are kind of sudden geohazards and mainly induced by abrupt changes of hydrodynamic conditions within karst conduit systems. Traditionally, only groundwater level monitoring was applied to monitoring the karst collapses, which did not reflect the hydrodynamic conditions within karst conduit systems. The monitoring technology of groundwater-air pressure was proposed by the Institute of Karst Geology, China in 1998 to study the formation mechanisms, monitoring and forecast of karst collapses. This technology has been improved in the past 20 years and applied to collapse risk assessment, triggering factor evaluation, and controlling drawdown of the groundwater levels during underground civil engineering works at 11 study sites. This advanced technology is characterized by improved borehole drilling, special sealing technique of borehole orifice, and high-frequency data logging. The recommended logging interval is at least 5–20 min to capture the transient changes of groundwater-air pressure within karst conduit systems. The pressure transducers used in this technique are retrievable and can be reused or recycled (Jiang et al., 2019a).

该研究总结了岩溶塌陷地下水 – 气压力监测技术。岩溶塌陷往往在没有预警的情况下迅速发生，潜在岩溶塌陷的早期预警一直是世界上最具挑战性的问题之一。事实上，大部分岩溶塌陷是突发性的地质灾害，主要是由岩溶管道系统内水动力条件的突变引起的。传统岩溶塌陷监测技术仅采用地下水位监测，不能反映岩溶管道系统内的水动力条件。早在1998年，中心科研团队就提出了地下水 – 气压力监测技术，旨在研究岩溶塌陷形成机制并开展监测和预警。经过20年的不断改进，这项先进技术应用于11个研究工作点地下土木工程建设中的塌陷风险评估、触发因素评估和地下水位下降防控。这项先进技术的特点是改进了钻孔技术、特殊的钻孔孔口密封技术和高频数据测井。建议的测井间隔至少为5~20分钟，以捕捉岩溶管道系统内地下水 – 气压力的瞬态变化。该项技术使用的压力传感器是可循环利用的，可重复使用或回收（Jiang et al., 2019a）。

Application examples of karst groundwater-air pressure monitoring technology
岩溶地下水－气压力监测技术应用实例

ID 编号	Study site 应用场地	City 城市	Province 省	Date of sinkhole occurrence 塌陷日期	Encountered problems 面临问题	Monitored since 监测起始年	Duration 监测持续时间	Number of monitoring wells 监测点个数	Object served 服务对象
1	Zhemu 柘木	Guilin 桂林	Guangxi 广西	1997/11	Sinkholes 岩溶塌陷	2000	17 years 17 年	6	Council 市政
2	YingHong 英红	Yingde 英德	Guangdong 广东		Potential sinkhole occurrence 岩溶塌陷隐患	2004	3 years 3 年	6	Wu-Guang HSR 武广高铁桥梁
3	Mingjiashan 名甲山	Anshan 鞍山	Liaoning 辽宁		Potential sinkhole occurrence 岩溶塌陷隐患	2008	1 mouth 1 个月	8	Ha-Da HSR 哈大高铁桥梁
4	Gongcheng 恭城	Guilin 桂林	Guangxi 广西		Potential sinkhole occurrence 岩溶塌陷隐患	2007	4 mouths 4 个月	10	Gui-Guang HSR 贵广高铁路基
5	Jinshazhou 金沙洲	Guangzhou 广州	Guangdong 广东	2006	Sinkholes 岩溶塌陷	2007	10 years 10 年	18	Wu-Guang HSR Tunnel 武广高铁隧道
6	Jiuxian 旧县	Taian 泰安	Shandong 山东	1988	Sinkholes 岩溶塌陷	2012	1 year 1 年	3	Well field 水源地
7	Qingyun 青云	Guigang 贵港	Guangxi 广西	1980	Sinkholes 岩溶塌陷	2012	5 years 5 年	1	Gas pipeline 油气管线
8	Jili 吉利	Laibin 来宾	Guangxi 广西	2010/6/3	Sinkholes 岩溶塌陷	2013	4 years 4 年	13	Nan-Liu HSR 南柳高铁路基
9	Dachengqiao 大成桥	Ningxiang 宁乡	Hunan 湖南	1982	Sinkholes 岩溶塌陷	2013	4 years 4 年	15	Mining 矿区
10	Chaoshan 朝山	Tongling 铜陵	Anhui 安徽	1982	Sinkholes 岩溶塌陷	2014	3 years 3 年	10	Mining 矿区
11	Zhongliangshan 中梁山	Chongqing 重庆	Chongqing 重庆	2015	Sinkholes 岩溶塌陷	2016	1 year 1 年	4	Tunnel 隧道

2. Hydraulic fracturing effect on punching-induced cover-collapse sinkholes: A case study in Guangzhou, China

钻孔引致覆盖层塌陷的水力压裂效应——以中国广州为例

Punching-induced cover-collapse sinkholes occur rapidly, and multiple collapses occur simultaneously compared with those caused by other common factors, such as pumping, rainstorms and mining. Erosion cannot explain the formation of this type of collapse entirely. We studied the hydraulic fracturing effect from geological analysis, field groundwater monitoring, laboratory testing, and pressure calculation. Four stages can be established for the development of a cover-collapse sinkhole based on a case study. Hydraulic fracturing and the suction effect cause sinkhole opening with the percussion bit falling and pulling repeatedly when the pile foundation is being built. The behavior can be likened figuratively to a piston action. The pressure of the calculated hydraulic fracturing effect is 6.05×10^4 Pa, which is equivalent to a 5.93-m-high water column and is much bigger than the critical pressure of 0.7 m at which hydraulic fracturing occurs in laboratory testing. The significance of this study from a practical perspective is to provide a preventative, early warning method for sinkhole opening by groundwater monitoring during pile foundation construction (Meng et al., 2020).

与其他常见因素（如抽水、暴雨和采矿）相比，钻孔引起的覆盖层塌陷发生迅速，且可同时发生多次塌陷，侵蚀不能完全解释这种塌陷的成因。本研究通过地质条件分析、现场地下水监测、实验室测试和压力计算等开展了水力压裂效果研究。通过实例分析，可以确定盖层塌陷发育的四个阶段。在桩基施工过程中，由于水力压裂和抽吸效应，冲击钻头反复下落和拔动，造成开放塌陷坑，这种行为类似活塞运动。计算出的水力压裂压力效应为 6.05×10^4 Pa，相当于一个 5.93 m 高的水柱，远大于实验室测试中水力压裂发生时 0.7 m 的临界压力。本研究旨在通过地下水监测为桩基施工过程中可能出现的开放性塌陷坑提供早期预警（Meng et al., 2020）。

Chapter 3 Scientific Research

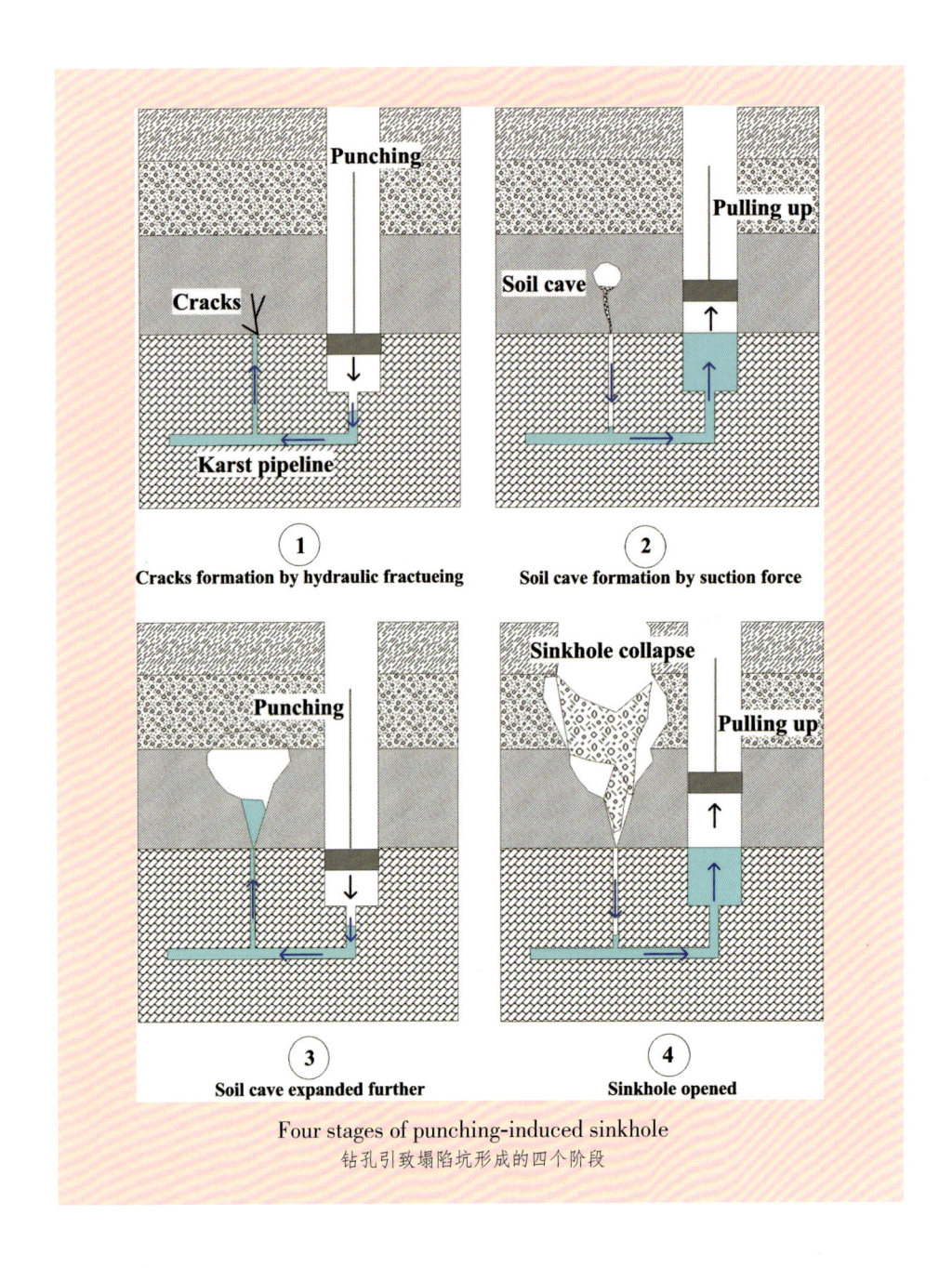

1. Cracks formation by hydraulic fractueing
2. Soil cave formation by suction force
3. Soil cave expanded further
4. Sinkhole opened

Four stages of punching-induced sinkhole
钻孔引致塌陷坑形成的四个阶段

3. A multidisciplinary approach in cover-collapse sinkhole analyses in the mantle karst from Guangzhou City (SE China)

广州覆盖型岩溶区盖层塌陷多学科分析（中国东南部）

Mantled karst is characterized by solution processes in underground conditions, where karst evidence can be masked. In mantled karst environments, a cover-collapse sinkhole can be considered one of the most hazardous elements. This study seeks to develop a methodological framework utilizing different techniques and approaches to understand cover-collapse sinkhole genesis and its likely evolution. The study area is located in the Conghua District of Guangzhou City, Guangdong Province, in the southeastern region of China. A mapping procedure was introduced to combine data from aerial photographs and intensive field investigations. In addition, data interpretations from borehole drilling activities and different geophysical approaches were performed to reconstruct the Quaternary deposit features, rock head morphology, and karst features. During these investigations, the detailed typology, morphometry, and chronology inventory of 49 cover-collapse sinkholes were analyzed, and three karst fissure zones covered by Quaternary soil were observed. In the study, the hydrogeological data suggested that karst aquifer pumping triggered the drop in groundwater levels. Cover-collapse sinkholes might be ascribed to erosion of the soil layers due to the groundwater level decline (Jia et al., 2021).

> 覆盖型岩溶以地下溶蚀过程为特征，其岩溶发育的事实通常被掩盖。在覆盖型岩溶环境中，盖层塌陷被认为是最危险的因素之一。本研究尝试利用不同的技术和方法来建立一个方法论框架，以探索盖层塌陷的成因及其可能的演变。研究区位于中国东南部的广东省广州市从化区。本研究通过映射过程将航空照片和高密度的实地调查数据相结合。此外，通过钻孔和不同的地球物理方法对数据进行解译，以重建第四纪以来的沉积特征、岩石形态和岩溶特征。调查过程中，对49个盖层塌陷坑进行了详细的类型学、形态计量学和年代学分析，观测到3个第四纪土壤覆盖的岩溶裂隙带。在本研究中，水文地质数据表明，岩溶含水层的抽水引发了地下水位的下降。盖层塌陷坑的形成可能由地下水位下降对土层的侵蚀引起（Jia et al., 2021）。

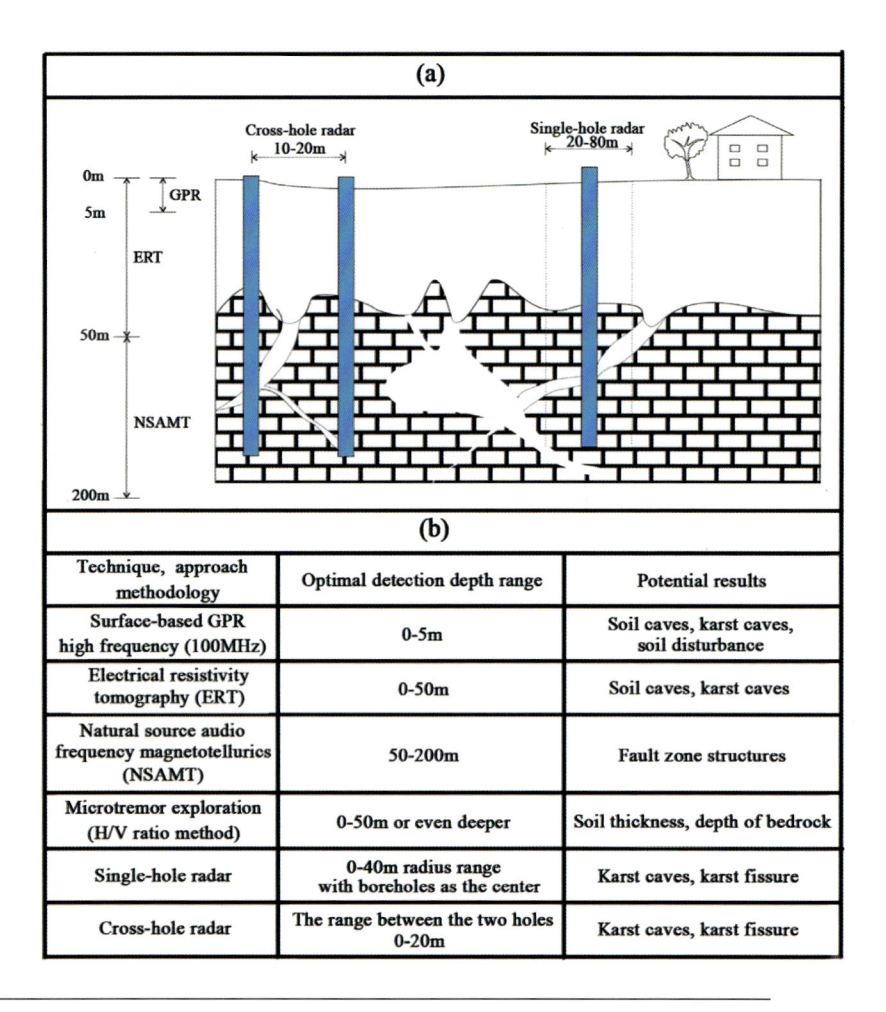

Technique, approach methodology	Optimal detection depth range	Potential results
Surface-based GPR high frequency (100MHz)	0-5m	Soil caves, karst caves, soil disturbance
Electrical resistivity tomography (ERT)	0-50m	Soil caves, karst caves
Natural source audio frequency magnetotellurics (NSAMT)	50-200m	Fault zone structures
Microtremor exploration (H/V ratio method)	0-50m or even deeper	Soil thickness, depth of bedrock
Single-hole radar	0-40m radius range with boreholes as the center	Karst caves, karst fissure
Cross-hole radar	The range between the two holes 0-20m	Karst caves, karst fissure

Integrated analysis of the different techniques and approaches: (a) Diagrammatic sketch of geophysical method framework in cover-collapse sinkhole analyses; (b) Characterization of the different techniques and approaches, a table with the optimal detection depth range of each technique and potential results

不同技术方法综合分析：(a) 覆盖层塌陷分析中的地球物理方法示意图；(b) 不同技术方法特征，各技术最佳的探测深度范围和推测性结果列表

4. Mechanism of sinkhole formation during groundwater-level recovery in karst mining area, Dachengqiao, Hunan Province, China
湖南大成桥岩溶矿区地下水位恢复过程中岩溶塌陷的形成机理

The Meitanba Coal Mine area in Hunan Province, China, had been impacted by severe cover collapse sinkholes since 1982 due to mine dewatering. After the coal mine was closed in February 2015, the groundwater level has increased significantly. A series of sinkholes were recorded in the study area during groundwater-level recovery. Analysis of monitoring results and in-situ investigation indicated that 13 sinkhole collapses were more likely induced by abrupt change of groundwater-air pressure in response to heavy rainfall from March 2015 to July 2016 when the groundwater level increased by as much as 76 m. When the karst conduit was flooded, a relatively sealed environment was formed between saturated sediments and flooded karst conduit. The implosion of entrapped air might have caused the cave roof to collapse followed by surface collapses in a short time. On the other hand, four sinkholes occurred in November 2016 when the groundwater levels were near the soil-bedrock interface at elevations between 52.5 and 58.9 m amsl, and the groundwater-level increase was at slower paces. Field measurements indicate that the groundwater-level fluctuation at the soil-bedrock interface could enlarge the soil cavity and accelerate the subsoil erosion process (Pan et al., 2018).

自 1982 年以来，由于采矿疏干影响，湖南煤炭坝矿区周边发生了大量岩溶塌陷现象。在 2015 年 2 月闭矿后，煤炭坝周边地下水位开始大幅上升，同时大成桥地区内发生了多起岩溶塌陷事件。监测结果分析和现场调查发现，自 2015 年 3 月至 2016 年 7 月，在降雨补给下大成桥地区地下水位升幅达 76 m，由此产生的 13 起岩溶塌陷与地下水 – 气压力急剧变化关系密切。其形成机理为：当岩溶管道充满水时，饱水盖层与岩溶管道水位之间形成了一个相对封闭的环境，当地下水位上升时，压缩管道内部气体并对上覆盖层产生气爆效应，导致盖层顶板破坏并在短时间内引起地面塌陷。另外，在 2016 年 11 月内研究区发生的 4 起岩溶塌陷均处于地下水位在岩土界面附近缓慢波动期间（水位标高为 52.5~58.9 m）。野外实测表明，地下水在基岩面附近波动，加快了土体的崩解并促进了土洞向地表发育，最终导致塌陷形成（Pan et al., 2018）。

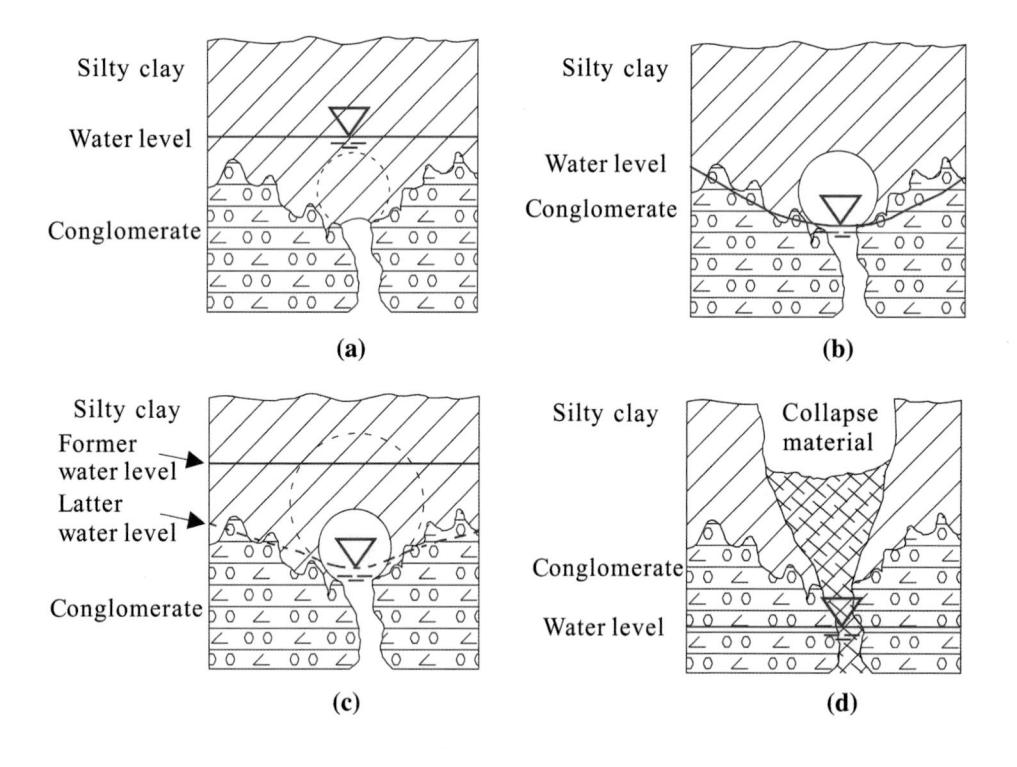

Schematic drawing of subsoil erosion in sinkhole development. (a) Alluvial deposits are saturated when groundwater level rises above the soil-bedrock interface. (b) Soil cavity forms overlain the bedrock when groundwater level declines. (c) Soil cavity begins to propagate upwards, when the groundwater level fluctuates near soil-bed-rock interface. (d) Collapse sinkholes occur on the surface

岩溶塌陷发育过程中土洞演化示意图：(a) 当地下水位在岩土界面以上时，覆盖层土体吸水饱和；(b) 当地下水位下降时，岩土界面附近形成土洞；(c) 当地下水位在岩土界面附近波动时，土洞开始向地表发育；(d) 地表出现塌陷坑

5. AHP-based evaluation of the karst collapse susceptibility in Tailai Basin, Shandong Province, China
基于 AHP 的山东泰莱盆地岩溶塌陷易发性评价

Karst sinkholes began to appear in Tailai Basin of China in the 1970s and have been increasingly exacerbating ever since. Evaluation of the sinkhole susceptibility has been challenging. The lithology (Li), distance to fault (DF), overburden type (OTy), overburden structure (OS), overburden thickness (OTh), groundwater depth (GD), relative location between the groundwater table and bedrock (RL), pumping wells (PW) are chosen as conditioning factors. The weights of the factors are obtained based on the analytic hierarchy process (AHP). Further, a GIS-based evaluation of the karst sinkholes susceptibility in the Tailai Basin is made. The Tailai Basin is divided into extremely susceptible area, highly susceptible area, moderately susceptible area, and weakly susceptible area, and their corresponding areas are 60.15 km^2, 85.80 km^2, 61.69 km^2 and 19.88 km^2, respectively. The extremely and highly susceptible areas are mainly distributed in Taian City and encompass an area of 145.95 km^2. The results of this study provide a decision basis for the prevention and treatment of karst sinkholes in the Tailai Basin (Wu et al., 2018).

20 世纪 70 年代以来，中国泰莱盆地开始出现岩溶塌陷，并不断加剧。岩溶塌陷易发性评价一直具有挑战性。选择岩性（Li）、距断层距离（DF）、覆盖层类型（OTy）、覆盖层结构（OS）、覆盖层厚度（OTh）、地下水波动幅度（GD）、地下水位与基岩面的相对位置（RL）、抽水井开采强度（PW）作为评价因子，并基于层次分析法（AHP）得到各评价因子的权重。进一步，借助 GIS 软件对泰莱盆地进行岩溶塌陷易发性评价。评价结果显示，泰莱盆地可分为极高易发区、高易发区、中等易发区和低易发区，其对应面积分别为 60.15 km^2、85.80 km^2、61.69 km^2 和 19.88 km^2。极高和高度易发区主要分布在泰安市，面积 145.95 km^2。研究结果为泰莱盆地岩溶塌陷地质防治工作提供了决策依据（Wu et al., 2018）。

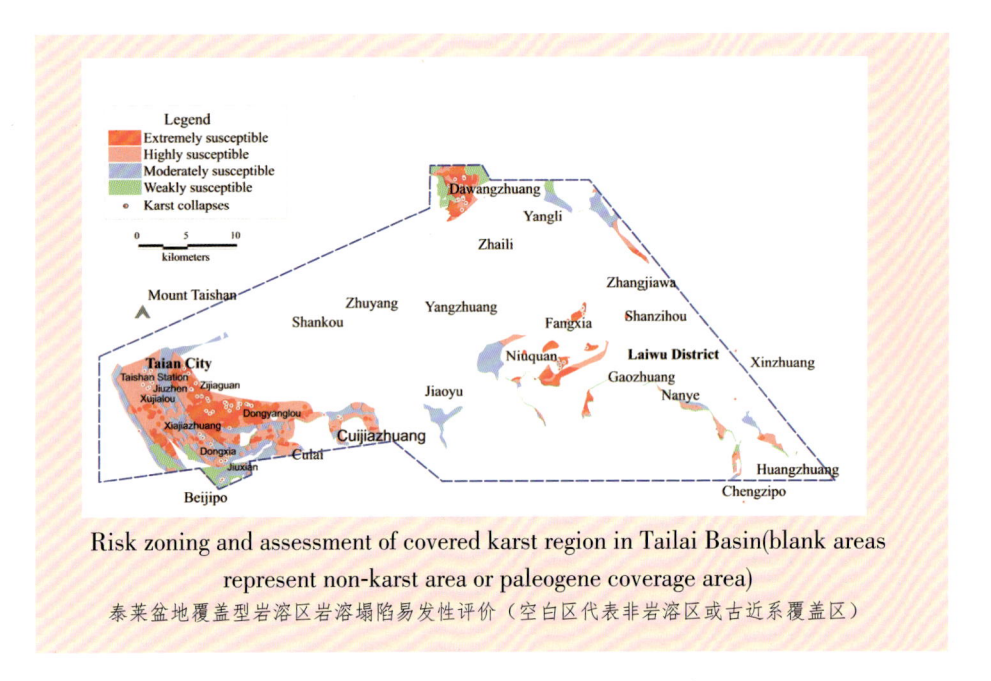

Risk zoning and assessment of covered karst region in Tailai Basin(blank areas represent non-karst area or paleogene coverage area)

泰莱盆地覆盖型岩溶区岩溶塌陷易发性评价（空白区代表非岩溶区或古近系覆盖区）

6. Application of seismic velocity tomography in investigation of karst collapse hazards, Guangzhou, China
地震速度层析成像法在中国广州岩溶塌陷灾害调查中的应用

The study applied seismic velocity tomography to the investigation and assessment of karst collapse hazards to facilitate accurate characterization of geological conditions of karst sinkhole formation. In the survey areas of Xiamao, Guangzhou, China, and Huangqi, Foshan, China, seismic velocity tomography was used to explore the structures of rock and soil associated with karst collapse. The results show that sand intercalated with clay or clay intercalated with soft soil dominates the cover of these two areas. The overburden is 20–33 m thick and underlain by Carboniferous limestone. In the limestone, there are well-developed karst caves and cracks as well as highly fluctuating bedrock surfaces. The seismic velocities are less than 2,500 m/s in the cover, 2,500–4,500 m/s in the karst fracture zones and caves of Xiamao, and 1,500−2,000 m/s in the Huangchi collapse area. The karst fracture zones, relief of bedrock surfaces, and variations of soil thicknesses revealed by seismic velocity tomography are well constrained and in agreement with those in the drilling borehole profiles. This study demonstrates that seismic velocity tomography can delineate anomalies of rock and soil with the advantages of speed, intuitive images, and high resolution (Zhao et al., 2018) .

本研究将地震速度层析成像技术应用在岩溶塌陷危险性调查和评价中，以便于精准掌握岩溶塌陷形成的地质条件。在中国广州夏茅和佛山黄岐调查区，利用地震速度层析成像技术探测了与岩溶塌陷有关的岩土结构。结果表明，砂夹黏土或黏土夹软土在这两个地区的覆盖层中占主导地位。覆盖层厚度为20~30 m，下伏石炭系灰岩。灰岩中可见发育良好的溶洞和裂缝，基岩表面起伏不平。覆盖层的地震速度小于2500 m/s，夏茅岩溶断裂带和洞穴的地震速度为2500~4500 m/s，黄岐塌陷区的地震速度为1500-2000 m/s。地震速度层析成像在揭示岩溶断裂带、基岩表面起伏和土壤厚度变化方面应用效果良好，其结果与钻孔剖面一致。本研究表明，地震速度层析成像具有速度快、图像直观、分辨率高等优点，可反演岩层和土层的异常（Zhao et al., 2018）。

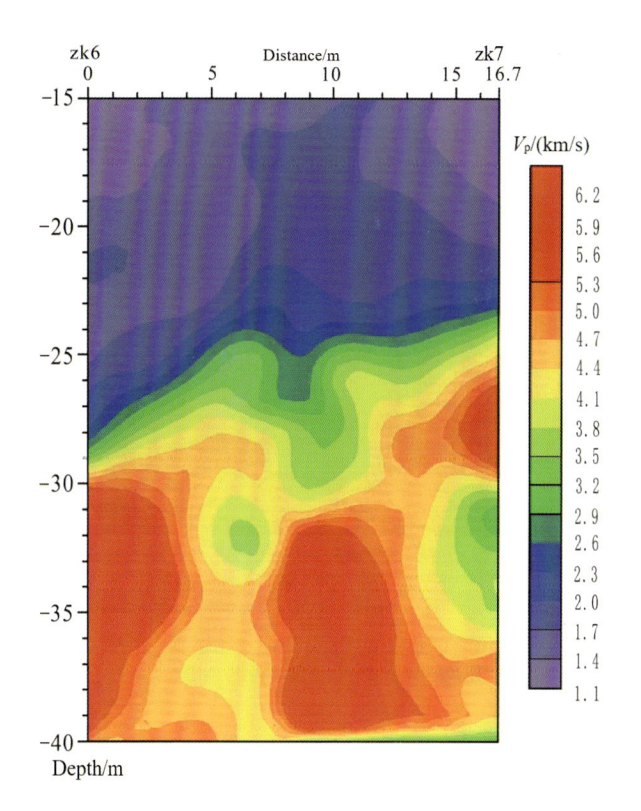

Seismic velocity model for profile zk6-zk7
zk6-zk7 剖面的地震速度模型

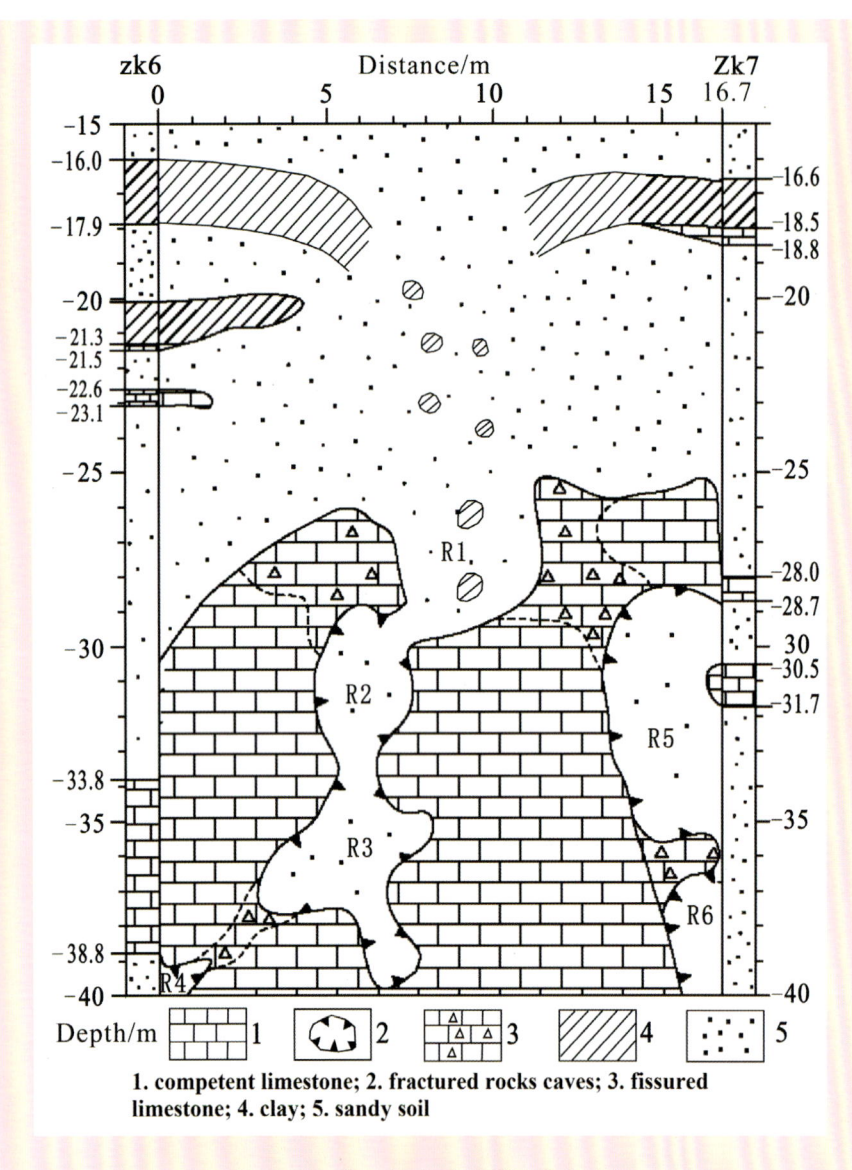

Interpreted geological profile zk6-zk7
zk6-zk7 解译后的地质剖面

1. competent limestone; 2. fractured rocks caves; 3. fissured limestone; 4. clay; 5. sandy soil

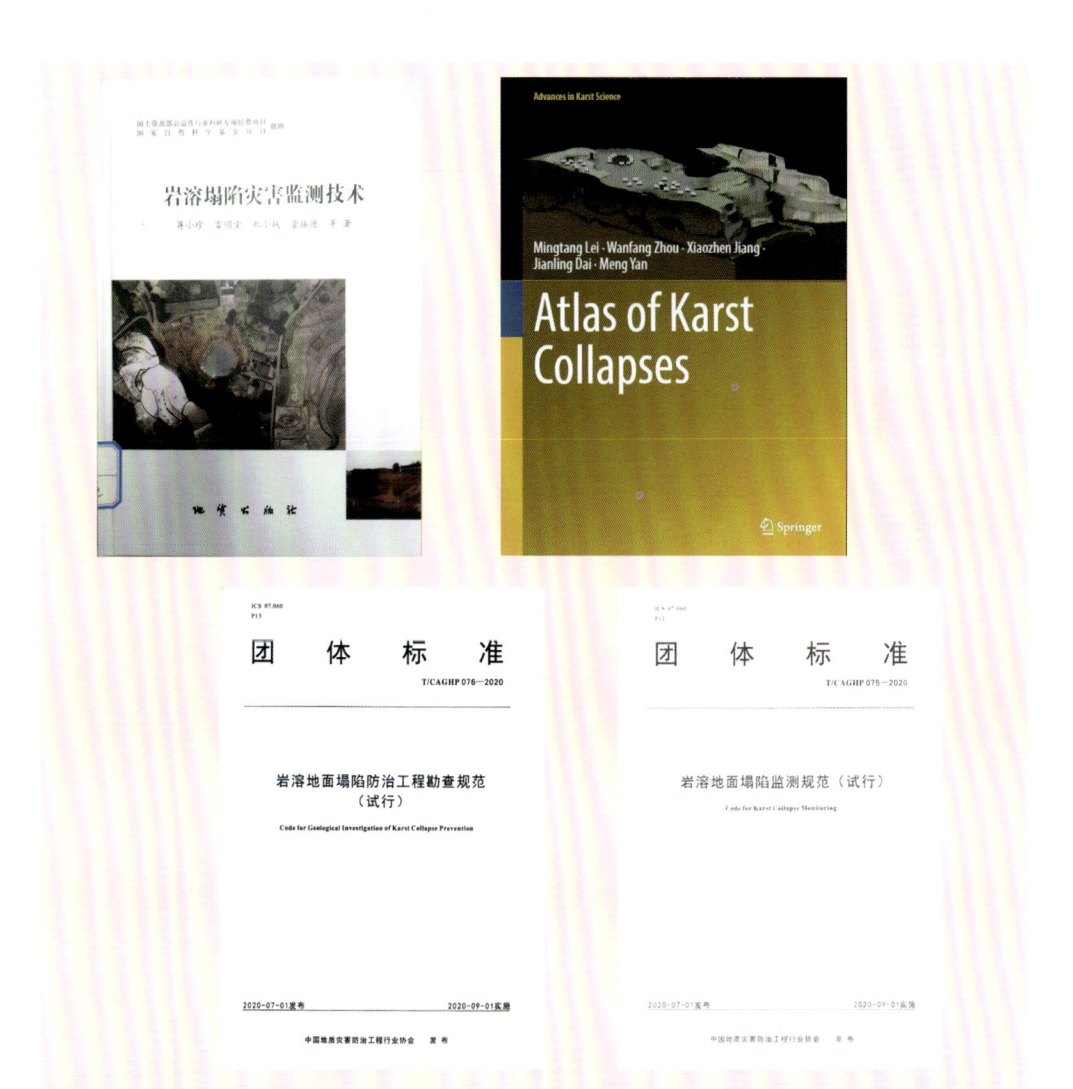

Published 2 monographs, compile 2 group standards: *Karst Collapses Monitoring Technology*; *Atlas of Karst Collapses*; *Code for Geological Investigation of Karst Collapse Prevention*; *Code for Karst Collapse Monitoring*

出版著作 2 部，编写团体标准 2 份：《岩溶塌陷灾害监测技术》《岩溶塌陷图集》《岩溶地面塌陷防治工程勘查规范（试行）》《岩溶地面塌陷监测规范（试行）》

3.2.6 Serial maps on karst compiled by IRCK
岩溶编图系列成果

During the second phase of operation, based on international geological survey projects, taking relevant countries in Southeast Asia as key studying areas, IRCK has prepared 28 karst-related serial maps at global, regional, and provincial scales. The maps include the *Karst Map of the World (1:10 million)*, *Serial Maps of Karst Environmental Geology in Southern China and Southeast Asia*, the *Map of Geological Disaster Prone Areas in Key Areas of the Five Countries on Indo-China Peninsula*, *Hydrogeological Map of Cambodia (1:500,000)*, and *Natural Landscape Resources Atlas of Guangxi Zhuang Autonomous Region*. IRCK participated in the preparation of the *World Karst Aquifer Map* supported by the International Association of Hydrogeologists (IAH) and the International Hydrological Programme (IHP), providing important scientific and technological support for the sustainable utilization of karst resources and the site selection of major constructions in different countries.

中心二期运营期内，依托境外地质调查项目，以东南亚相关国家为重点工作区，编制了28幅全球尺度、区域尺度及省市级尺度的岩溶相关系列图，如《全球岩溶分布图（1:1000万）》《中国南部及东南亚地区岩溶环境地质系列图》《中南半岛5国重点区地质灾害易发区分布图》《柬埔寨1:50万水文地质图》《广西壮族自治区自然景观资源图集》等，参与编制由国际水文地质学家协会及政府间水文计划（IHP）组织支持的《世界岩溶含水层图》，为各国岩溶资源可持续利用、重大工程选址等提供了重要科技支撑。

上图 / Top

Serial Maps of Karst Environmental Geology in Southern China and Southeast Asia

《中国南部及东南亚地区岩溶环境地质系列图》

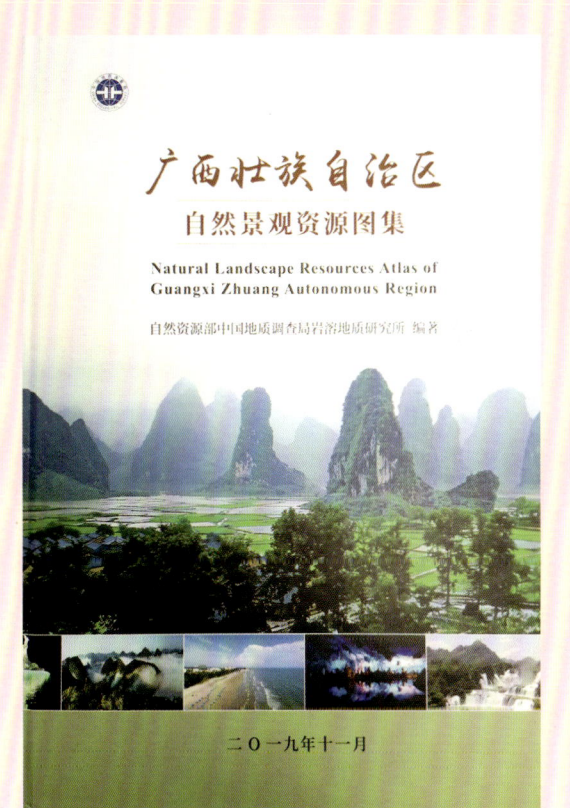

下图 / Bottom

Natural Landscape Resources Atlas of Guangxi Zhuang Autonomous Region

《广西壮族自治区自然景观资源图集》

参考文献

曹建华, 邓艳, 杨慧, 等. 2016. 喀斯特断陷盆地石漠化演变及治理技术与示范. 生态学报, 36(22): 7103–7108.

曹建华, 蒋忠诚, 袁道先, 等. 2023. 中国西南岩溶碳循环及全球意义. 北京: 测绘出版社: 177–179.

邓艳, 蒋忠诚, 徐烨, 等. 2018. 典型表层岩溶泉域植被对降雨的再分配研究. 中国岩溶, 37(5): 714–721.

黄芬. 2020. 漓江流域氮素对岩溶碳循环过程的影响机制. 北京: 中国地质科学院.

蒋忠诚, 罗为群, 童立强, 等. 2016. 21世纪西南岩溶石漠化演变特点及影响因素. 中国岩溶, 35(5): 461–468.

蒋忠诚, 张晶, 黄超, 等. 2019. 湘西地质公园岩溶峡谷群成因及其地学意义. 中国岩溶, 38(2): 269–275.

蓝芙宁, 张贵, 张华, 等. 2021. 喀斯特断陷盆地石漠化演变及综合治理技术与示范科技报告.

李文莉, 蒋忠诚, 张冉. 2021. 中国地质调查局西南岩溶石漠化野外科学观测研究站简介. 中国地质, 48(1): 345–346.

梁永平, 申豪勇, 赵春红, 等. 2021. 对中国北方岩溶水研究方向的思考与实践. 中国岩溶, 40(3): 363–380.

潘晓东, 曾洁, 任坤, 等. 2018. 贵州毕节岩溶斜坡地带地下水赋存规律与钻探成井模式. 地球学报, 39(5): 606–612.

王赛男, 蒲俊兵, 李建鸿, 等. 2019. 岩溶断陷盆地"盆-山"耦合地形影响下的气候特征及其对石漠化生态恢复的影响探讨. 中国岩溶, 38(1): 50–59.

王宇, 张华, 张贵, 等. 2017. 喀斯特断陷盆地环境地质分区及功能. 中国岩溶, 36(3): 283–295.

韦跃龙, 陈伟海, 罗劬侃. 2018. 贵州织金洞世界地质公园喀斯特成景机制及模式研究. 地质论评, 64(2): 457–476.

杨杨, 赵良杰, 苏春田, 等. 2019. 基于CFP的岩溶管道流溶质运移数值模拟研究. 水文地质工程地质, 46(4): 51–57.

袁道先, 蔡桂鸿. 1988. 岩溶环境学. 重庆: 重庆出版社: 128.

曾锐, 张陶, 蒲俊兵, 等. 2020. 典型岩溶断陷盆地农作物生态需水研究——以蒙自地区为例. 中国岩溶, 39(6): 873–882.

赵良杰, 王莹, 周妍, 等. 2022. 基于SWAT模型的珠江流域地下水资源评价研究. 地球科学.

Bai B, Jiang Z C, Zhang C, et al. 2023. New achievements of IGCP 661 structure, substance cycle, and environment sustainability of the critical zone in karst systems (2017–2021). Episodes.

Cao J H, Bill H, Groves C, et al. 2016. Karst dynamic system and the carbon cycle. Zeitschrift für Geomorphologie, 60(Suppl. 2): 35–55.

Dodge-Wan D, Prasanna M V, Nagarajan R, et al. 2017. Epiphreatic caves in Niah karst tower (NW Borneo): Occurrence, morphology and hydrogeochemistry. Acta Carsologica, 46(2–3): 149–163.

Ford D, Williams P. 2007. Karst Hydrogeology and Geomorphology. Chichester, UK: Wiley: 5.

Gan F P, Han K, Lan F N, et al. 2017. Multi-geophysical approaches to detect karst channels underground—A case study in Mengzi of Yunnan Province, China. Journal of Applied Geophysics, 136: 91–98.

Garousi F, Shan Z J, Ni K, et al. 2021. Decreased inorganic N supply capacity and turnover in calcareous soil under degraded rubber plantation in the tropical karst region. Geoderma, 381: 114754.

Guo F, Jiang G H, Zhao H L, et al. 2019. Physicochemical parameters and phytoplankton as indicators of the aquatic environment in karstic springs of South China. Science of the Total Environment, 659: 74–83.

Guo Y L, Zhang W, Zhang C, et al. 2021. Contamination characteristics of chlorinated hydrocarbons in a fractured karst aquifer using TMVOC and hydro-chemical techniques. Science of the Total Environment, 794: 148717.

Huang F, Vasic L, Wu X, et al. 2019. Hydrochemical features and their controlling factors in the Kucaj-Beljanica Massif, Serbia. Environmental Earth Sciences, 78: 498.

Jia L, Meng Y, Li L J, et al. 2021. A multidisciplinary approach in cover-collapse sinkhole analyses in the mantle karst from Guangzhou City (SE China). Natural Hazards, 108: 1389–1410.

Jiang G H, Chen Z, Siripornpibul C, et al. 2021. The karst water environment in Southeast Asia: Characteristics, challenges, and approaches. Hydrogeology Journal, 29: 123–135.

Jiang X Z, Lei M T, Zhao H Q, et al. 2019a. Review of the advanced monitoring technology of groundwater-air pressure (enclosed potentiometric) for karst collapse studies. Environmental Earth Sciences, 78: 701.

Jiang Z C, Zhang C, Qin X Q, et al. 2019b. Structural features and function of the karst critical zone. Acta Geologica Sinica (English Edition), 93(supp. 1): 109–112.

Li J H, Pu J B, Zhang T. 2022. Transport and transformation of dissolved inorganic carbon in a subtropical groundwater-fed reservoir, south China. Water Research, 209: 117905.

Li Q, Song A, Yang H, et al. 2021. Impact of rocky desertification control on soil bacterial community in karst graben basin, Southwestern China. Frontiers in Microbiology, 12: 636405.

Meng Y, Jia L, Huang J M, et al. 2020. Hydraulic fracturing effect on punching-induced cover-collapse sinkholes: A case study in Guangzhou, China. Arabian Journal of Geosciences, 13: 28.

Pan Z Y, Jiang X Z, Lei M T, et al. 2018. Mechanism of sinkhole formation during groundwater-level recovery in karst mining area, Dachengqiao, Hunan Province, China. Environmental Earth Sciences, 77: 799.

Sun P A, He S Y, Yu S, et al. 2021. Dynamics in riverine inorganic and organic carbon based on carbonate weathering coupled with aquatic photosynthesis in a karst catchment, Southwest China. Water Research, 189: 116658.

Urushibara-Yoshino K. 1991. Landuse and soils in karst areas of Java Indonesia//Sauro U, Bondesan A, Meneghel M. Proceedings of the International Conference on Environmental Changes in Karst Areas. IGU, UIS, Italy: 61–67.

Uysal I T, Ünal-Imer E, Shulmeister J, et al. 2019. Linking CO_2 degassing in active fault zones to long-term changes in water balance and surface water circulation, an example from SW Turkey. Quaternary Science Reviews, 214: 164–177.

Wang P, Hu G, Cao J H. 2017. Stable carbon isotopic composition of submerged plants living in karst water and its eco-environmental importance. Aquatic Botany, 140: 78–83.

Wang Z J, Yin J J, Cheng H, et al. 2022. Climatic controls on travertine deposition in southern Tibet during the late Quaternary. Palaeogeography, Palaeoclimatology, Palaeoecology, 589: 110852.

Wu L J, Jiang H X, Chen W H, et al. 2021. Geodiversity, geotourism, geoconservation, and sustainable development in Xiangxi UNESCO Global Geopark—A case study in ethnic minority areas. Geoheritage, 13: 99.

Wu Y B, Jiang X Z, Guan Z D, et al. 2018. AHP-based evaluation of the karst collapse susceptibility in Tailai Basin, Shandong Province, China. Environmental Earth Sciences, 77: 436.

Yang H, Prelovšek M, Huang F, et al. 2019a. Quantification and evaluation of soil organic carbon and its fractions: Case study from the classical karst, SW Slovenia. Acta Carsologica, 48/3: 295–311.

Yang H, Zhang P, Zhu T B, et al. 2019b. The characteristics of soil C, N, and P stoichiometric ratios as affected by geological background in a karst graben area, Southwest China. Forests, 10(7): 601.

Yin J J, Tang W, Wang Z J, et al. 2021. Deciphering the hydroclimatic significance of dripwater $\delta^{13}C_{DIC}$ variations in monsoonal China based on modern cave monitoring. Journal of Hydrology, 603: 126882.

Yuan D X. 1997. Rock desertification in the subtropical karst of south China. Zeitschrift für Geomorphologie, 108(Supp): 81–90.

Zhao W, Gan F P, Meng Y, et al. 2018. Application of seismic velocity tomography in investigation of karst collapse hazards, Guangzhou, China. Environmental Earth Sciences, 77: 258.

Zhu T B, Yang C, Wang J, et al. 2018. Bacterivore nematodes stimulate soil gross N transformation rates depending on their species. Biology and Fertility of Soils, 54: 107–118.

Name: Huanglong World Natural Heritage Site
Location: Sichuan
Inscribed as a UNESCO Natural World Heritage Site in 1992
Summary: Huanglong is renowned for its beautiful mountainous scenery, with relatively undisturbed and highly diverse forest ecosystems, combined with the more spectacular localized karst formations, such as travertine pools, waterfalls and limestone shoals. Its travertine terraces and lakes are certainly unique in all of Asia.

名称：黄龙世界自然遗产地
所在地点：四川
列入联合国世界遗产地时间：1992年
概述：黄龙以五彩斑斓的钙华池、光芒万丈的雪山、十步九曲的高山峡谷、神秘幽静的大森林著称于世，其中尤以高山彩湖、叠瀑为主的石灰华岩溶景观令人叹为观止。

Chapter 4

International Exchange and Training

第四章　国际交流与培训

From 2016 to 2021, IRCK has promoted multilateral and bilateral exchanges proactively through international conferences and annual international training courses, serving an efficient stage for experience sharing, and bringing about a totally new impetus to stimulate the brainstorm of karst researchers.

2016~2021年，中心通过积极举办、参与国际会议，组织年度国际培训等方式，大力推进多双边交流沟通，为全球岩溶科学的经验分享提供了高效舞台，为激发岩溶科研人员的头脑风暴带入了全新动力。

4.1 International and Domestic Conferences
国际国内会议

4.1.1 Organize important international and domestic conferences to create an international publicity stage for karst science
举办重要国际国内会议，打造国际岩溶学科宣讲舞台

As an important bridge to the karst world, IRCK has organized and co-organized various international and domestic conferences positively since 2016, creating a good academic exchange atmosphere for scholars from all over the world, and striving to build up an international promotion stage for karst science.

2016年以来，中心积极主办、协办各类国际国内会议，充分发挥桥梁作用，为世界各地学者营造了良好的学术交流氛围，力求打造国际岩溶学科宣讲舞台。

4.1.1.1 The events of the International Big Scientific Plan on "Global Karst"
"全球岩溶"国际大科学计划系列会议

1. The announcement of the International Big Scientific Plan on "Global Karst"
"全球岩溶动力系统资源环境效应"国际大科学计划启动会

IRCK hosted the announcement ceremony of the International Big Scientific Plan on "Resources and Environmental Effects of Global Karst Dynamic Systems" (Global Karst) in Guilin of China on November 14, 2016. Minister Jiang Daming of the Ministry of Land and Resources (MLR), together with Director-General Irina Bokova of UNESCO, sent congratulation letters respectively. Meanwhile, the leaders of Chinese government and representatives of international organizations attended the event, including president of China Geological Survey (CGS); the vice governor of Guangxi Zhuang Autonomous Region; the chief Engineer of MNR and the chairperson of the

Governing Board of IRCK; the specialist of UNESCO Beijing Office; the president of IUGS; and the director of Technical Secretariat of the Coordinating Committee for Geoscience Programmes in East and Southeast Asia (CCOP).

2016年11月14日,"全球岩溶动力系统资源环境效应"国际大科学计划(以下简称"全球岩溶"国际大科学计划)启动仪式在岩溶中心举行。中国国土资源部部长姜大明先生、联合国教科文组织总干事伊琳娜·博科娃分别致贺信;中国地质调查局局长,广西壮族自治区副主席,国土资源部总工程师、国际岩溶研究中心理事会主席,联合国教科文组织驻华代表处项目专员,国际地质科学联合会主席,东亚东南亚地学计划协调委员会技术秘书处主任等出席。

At the ceremony, IRCK released the *Report on Karst Geological Survey in China (2016)*, as well as the *Report on the Geological Survey on the Integrated Control of Karst Rocky Desertification in Southwest China (2016)*. Moreover, the Initiative of the International Big Scientific Plan on Global Karst was announced, appealing all the attendees to improve and implement Global Karst together, and contribute their wisdom to the reasonable use of resources and socio-economic development of karst areas.

会议发布了《中国岩溶地质调查报告(2016年)》《西南岩溶石漠化综合治理地质调查报告(2016年)》。会议宣读了"全球岩溶"国际大科学计划倡议书,呼吁与会科学家要共同协商,推进大科学计划的完善与实施,形成合力,为应对岩溶地区的资源合理利用和社会经济发展贡献岩溶地质科学家的智慧与才华。

In addition, as many as 12 scientist representatives from 11 countries like US and Brazil signed the Intent of Support for the Global Karst jointly, showing a common willingness to contribute their efforts for this big scientific plan.

来自美国、巴西等11个国家的12位科学家代表共同签署了"全球岩溶动力系统资源环境效应"支持函,体现了为大科学计划做出共同努力的意愿。

Intent of Support

大科学计划

Report on Karst Geological Survey in China (2016) and *Report on the Geological Survey on the Integrated Control of Karst Rocky Desertification in Southwest China (2016)*
《中国岩溶地质调查报告（2016年）》《西南岩溶石漠化综合治理地质调查报告（2016年）》

2. The Workshop on "Global Karst"
"全球岩溶"国际大科学计划专题研讨会

IRCK organized its workshop on "Global Karst" in Beijing on 11 July 2018, aiming to implement the guiding principles of the *Notice of the State Council on Printing and Distributing the Circular of Actively Leading and Organizing the International Big Scientific Plans and Programs*. IRCK hopes to carry out scientific and technological innovative cooperation with other countries, especially the partners along the Belt and Road, based on its academic advantages; meanwhile, it also hopes to support the international exchange and cooperation of Guilin, the National Innovative Demonstration Area for the Sustainable Development Agenda.

2018年7月11日,中心在北京组织召开"全球岩溶"国际大科学计划专题研讨会。会议旨在落实《国务院关于印发积极牵头组织国际大科学计划和大科学工程方案的通知》的精神,结合中国岩溶学科在国际上的学术优势,积极融入"一带一路"建设科技创新合作,带动桂林国家可持续发展议程创新示范区建设国际交流与合作研究。

Report in *Science and Technology Daily*
《科技日报》报道

Chaired by Prof. Jiang Zhongcheng (GB-II member of IRCK), the workshop has invited Prof. Cao Jianhua, the executive deputy director of IRCK to make a detailed introduction on the "Principles and Progress of the International Big Scientific Plan on Global Karst". All the attendees had a thorough discussion on the research plan and possible cooperative patterns, with a consensus reached on the following aspects: karst is a significant issue of earth science with common concerns globally; karst requires an international big scientific plan with wide cooperation basis; the experts of China has made outstanding achievement in karst research, being qualified to lead the joint research on karst; and karst science could play as an important support for Guilin to construct as a National Innovative Demonstration Area for the Sustainable Development Agenda.

会议由中心理事蒋忠诚主持，常务副主任曹建华做了"全球岩溶"国际大科学计划基础与进展的专题报告，与会专家就研究内容、国际合作方式进行了探讨。会议一致认为，岩溶是重要的地球科学问题，具有全球共识；岩溶需要国际大科学计划支持，具有国际合作基础；我国科学家在岩溶研究领域已处于国际领先地位，具备牵头和主导该领域研究与国际合作的能力；岩溶科学研究可以在桂林市国家可持续发展议程创新示范区建设中发挥重要科技支撑作用。

The meeting was attended by nearly 20 officials and experts from different organizations, including the Department of Science and Technology for Social Development of the Ministry of Science and Technology (MOST), the Administrative Centre for China's Agenda 21, the Department of Science and Technology and International Cooperation of the Ministry of Natural Resources (MNR), the Department of Science and Technology and International Cooperation of China Geological Survey (CGS), Department of Land and Resources of Guangxi Zhuang Autonomous Region, the College of Urban and Environmental Sciences of Peking University, and the Institute of Subtropical Agriculture of the Chinese Academy of Sciences, among others.

会议邀请中国科学技术部社会发展科技司、中国21世纪议程管理中心、自然资源部科技与国际合作司、中国地质调查局科技外事部、广西壮族自治区国土资源厅、北京大学城市与环境学院、中国科学院亚热带农业生态研究所等单位的近20位领导和专家出席。

Chapter 4 International Exchange and Training

上图/Top

Mr. Huang Jing, the director of the Administrative Centre for China's Agenda 21, gave a conclusion of the workshop
中国21世纪议程管理中心黄晶主任做总结发言

下图/Bottom

Mr. Cao Jianhua, the executive deputy director of IRCK, made a special report
岩溶中心曹建华常务副主任做专题汇报

3. The seminar of International Big Scientific Plan on "Global Karst"
"全球岩溶动力系统资源环境效应"国际大科学计划国际研讨会

IRCK held the seminar of International Big Scientific Plan on "Resources and Environmental Effects of Global Karst Dynamic Systems" (Global Karst) on 24 September 2019 in Guilin of Guangxi, with Mr. Li Pengde, member of the Standing Committee of CPPCC and vice president of CGS, attended and delivered a welcome speech.

2019年9月24日,"全球岩溶动力系统资源环境效应"国际大科学计划国际研讨会在广西桂林召开。全国政协常委、自然资源部中国地质调查局副局长李朋德出席会议并致辞。

The event invited experts from 22 countries, like Austria, Brazil, Russia, China, etc. They proposed instructive suggestions to Global Karst: it is necessary to make more efforts on the construction of karst critical zone monitoring network and comparative study; to acquire new understandings on carbon cycle and global climate change, rocky desertification control and ecological restoration, as well as landscapes and water resources development and utilization; to promote deep-earth and deep-time survey and research on karst, to focus on carbonate rocks environmental and temporal-spatial evolution, providing references for atmospheric evolution, continental drift, and life evolution; moreover, to find out the developing features of paleokarst in different areas, vertical zones, and different periods, providing technical support to the prospecting of solid minerals, oil, gas, and geothermal resources.

来自奥地利、巴西、俄罗斯、中国等22个国家的专家学者参加了本次会议。会上,专家表示"全球岩溶"国际大科学计划应着力推进全球岩溶关键带监测网站建设和对比研究,在岩溶碳循环与应对全球气候变化、石漠化治理与生态修复、景观资源–水资源开发利用方面力求取得新认识、新成果;推进岩溶地质调查研究向地球深部、深时发展,调查研究碳酸盐岩形成环境及时空演变,为认识地球大气演变、大陆漂移、生命演化提供科学数据;探索古岩溶在区域差异、垂向分带及时代分期上的发育特征,为固体矿产、石油天然气和地热资源的勘探开发提供技术支撑。

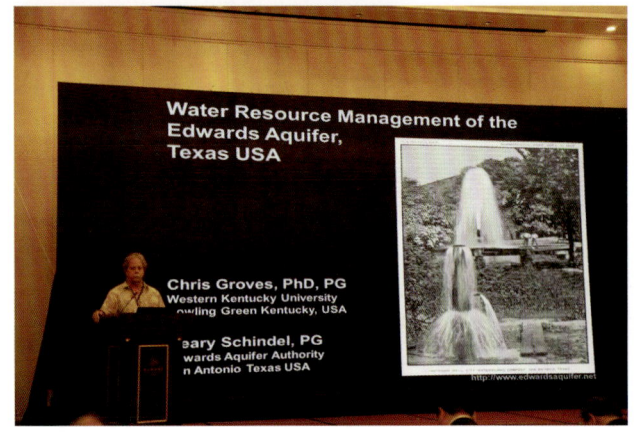

Chapter 4 International Exchange and Training

左上图 /Top left
Mr. Li Pengde, member of the Standing Committee of CPPCC and vice president of CGS, delivered a speech at the conference
全国政协常委、自然资源部中国地质调查局副局长李朋德为大会致辞

左下图 /Bottom left
Prof. Cheng Hai, the AC-II member of IRCK, introduced the progress of paleoclimate reconstruction through stalagmites
中心第二届学术委员会委员程海教授介绍石笋古气候重建成果

右上图 /Top right
Mr. Cao Jianhua, the executive deputy director of IRCK, introduced the progress of the "Global Karst"
中心曹建华副主任介绍"全球岩溶"国际大科学计划进展

右下图 /Bottom right
Prof. Chris Groves, the GB-II member of IRCK, introduced groundwater management of Edwards Aquifer
中心第二届理事会理事克里斯·葛立夫教授介绍美国得克萨斯州爱德华含水层水资源管理

The Second 6'Years of IRCK

4.1.1.2 ISO/TC 319 Karst 1st Plenary Meeting
国际标准化组织岩溶技术委员会会议

IRCK held the ISO/TC 319 Karst 1st Plenary Meeting successfully on 23-24 September 2019, with Mr. Stephane Savage, the Technical Programme Manager (TPM) from the ISO Central Secretariat (CS) attended, and more than 30 representatives from China, Austria, Canada, Saudi Arabia, Indonesia, Thailand, Serbia, Croatia, and South Africa attended as well. The delegation of China was composed of experts from the Development and Research Center of CGS, the Institute of Karst Geology of CGS/IRCK, the Institute of Earth Environment of Chinese Academy of Sciences (CAS), the Institute of Geochemistry of CAS, Nanjing University, Tianjin University, Southwest University, and Huazhong University of Science and Technology.

2019年9月23~24日，国际标准化组织岩溶技术委员会（ISO/TC 319）第一次全体会议顺利召开。国际标准化组织中央秘书处专员斯蒂芬·索瓦吉先生与来自中国、奥地利、加拿大、沙特阿拉伯、印度尼西亚、泰国、塞尔维亚、克罗地亚和南非等国的30多名代表参加了本次会议。中国代表团成员由中国地质调查局发展研究中心、中国地质科学院岩溶地质研究所、中国科学院地球环境研究所、中国科学院地球化学研究所、南京大学、天津大学、西南大学、华中科技大学等单位专家组成。

With a grand unveiling ceremony, the meeting discussed ISO/TC 319 *Strategic Commercial Prospectus* and set up a Chairman's Advisory Group; the meeting confirmed that the International Union of Speleology (UIS) would be a liaison; meanwhile, two new project proposals *Monitoring Technology for Karst Critical Zones* and *Terminology for Karst Caves* initiated by IRCK experts were introduced to all the attendees, indicating ISO/TC 319 Karst started its work on promotion of international standardization on karst formally.

本次会议举办了隆重的揭牌仪式。会议着重讨论了ISO/TC 319《战略性商业计划书》，并成立了主席顾问组；认定国际洞穴联合会为技术委员会联络组织，同时，推介了由中心专家发起的两项国际标准新工作项目提案：《岩溶关键带监测技术》《岩溶洞穴术语》。这标志着岩溶技术委员会开始正式推动岩溶领域的国际标准化工作。

Chapter 4　International Exchange and Training

国际标准化组织岩溶技术委员会第一次全体会议
ISO/TC 319 Karst 1st Plenary Meeting

The ISO/TC 319 Karst was jointly unveiled by the representatives as follows: Mr. Li Pengde, member of the Standing Committee of CPPCC and vice president of CGS (the fourth from the left); Mr. Cui Gang, director of the Department of Standardization Innovation, the State Administration for Market Regulation (SAC) (the third from the right); Mr. Yuan Daoxian, the GB-II member of IRCK and the academician of CAS (the third from the left); Mr. Stephane Savage, TPM from ISO CS (the second from the right); Ms. Liu Haiyan, the representative of the Department of Science and Technology Development of the Ministry of Natural Resources (the second from the left); Mr. Hu Maoyan, director of IRCK (the first from the right), and Mr. Jiang Zhongcheng, the chairman of ISO/TC 319 and the GB-II member of IRCK (the first from the left).

全国政协常委、自然资源部中国地质调查局李朋德副局长（左4）、国家市场监督管理总局（国家标准委）崔钢司长（右3）、中心理事、中国科学院院士袁道先先生（左3）、国际标准化组织总部代表斯蒂芬·索瓦吉先生（右2）、自然资源部科技发展司代表刘海岩女士（左2）、岩溶中心主任胡茂焱先生（右1）、国际标准化组织岩溶技术委员会主席、中心理事蒋忠诚先生（左1）共同揭牌。

The Second 6 Years of IRCK

4.1.1.3 The events under the China-ASEAN Mining Cooperation Forum
中国 – 东盟矿业合作论坛岩溶主题系列研讨会

Since 2014, IRCK has organized 5 karst-related seminars (sub-forums) under the China-ASEAN Mining Cooperation Forum in 2014, 2015, 2018, 2019, and 2021 respectively. During the phase-II operation, the seminars (sub-forums) in 2018, 2019, and 2021 focused on topics as follows.

自 2014 年始，中心分别于 2014 年、2015 年、2018 年、2019 年、2021 年承办了 5 届中国 – 东盟矿业合作论坛岩溶主题系列研讨会（分论坛）。在中心运行的第二个周期内，中心分别承担了 2018 年、2019 年、2021 年的研讨会（分论坛），具体内容如下。

1. The Seminar on China-ASEAN Geosciences Cooperation in Karst Hydrogeology and Environmental Geology
中国 – 东盟岩溶水文环境地学合作研讨会

IRCK hosted the Seminar on China-ASEAN Geosciences Cooperation in Karst Hydrogeology and Environmental Geology on 16 November 2018, during the China-ASEAN Mining Cooperation Forum. The seminar invited Mr. Li Jinfa, the GB-II member of IRCK and the president of the China Geological Survey, to deliver a welcome speech; Mr. Yuan Daoxian, the GB-II member of IRCK, to give a keynote speech; Mr. Jiang Zhongcheng, the GB-II member of IRCK, to introduce the progress on the compilation of serial maps of global karst (1:10,000,000), attracting great attention from all the attendees. In addition, the seminar also invited six experts and scholars from the US, Thailand, Myanmar, Indonesia, and China to make presentations on global or regional karst hydrogeology and environmental geology.

2018 年 11 月 16 日，在中国 – 东盟矿业合作论坛召开期间，中心承办了中国 – 东盟岩溶水文环境地学合作研讨会。会议邀请了中心理事、中国地质调查局局长李金发致辞，中心理事、学术委员会主任袁道先院士做主旨发言；中心理事蒋忠诚研究员宣传并推介了全球岩溶分布系列图（1:1000 万）编制进展，获得了与会专家的高度关注；此外，会议还邀请了来自美国、泰国、缅甸、

Chapter 4　*International Exchange and Training*

印度尼西亚和中国的 6 位专家学者围绕全球或区域的岩溶水文地质、环境地质问题展开交流研讨。

The attendees were listening to the presentations
中国-东盟岩溶水文环境地学合作研讨会现场

2. The Seminar on China-ASEAN Natural Landscape Resources Atlas (2019)
2019年中国－东盟自然资源图集研讨会

IRCK hosted the Seminar on China-ASEAN Natural Landscape Resources Atlas on 15 November 2019, during the China-ASEAN Mining Cooperation Forum, with over 100 experts and guests from CGS, the Department of Natural Resources of Guangxi Zhuang Autonomous Region (Guangxi DNR), the geology and mineral prospecting agencies of Guangxi, and the Leye–Fengshan UGGp Administrative Committee, etc. The seminar was composed of three parts: achievements introduction made by Prof. Chen Weihai, the expert in karst landscapes from IRCK, sharing an overview of natural landscape resources in Guangxi and the progress for the atlas compilation; presenting ceremony of the atlas, during which, the *Natural Landscape Resources Atlas of Guangxi Zhuang Autonomous Region* was presented to Guangxi DNR as a gift by IRCK; and thematic presentations, made by 5 experts from China, France, and UK on the investigation, development, and protection of karst landscape resources.

2019年11月15日，在中国－东盟矿业合作论坛召开期间，中心承办了中国－东盟自然景观资源图集研讨会。来自中国地质调查局、广西自然资源厅、广西壮族自治区相关地勘单位、中国乐业－凤山世界地质公园管理委员会等机构的国内外嘉宾100余人参加了此次会议。本次会议分为成果介绍、成果赠送和专题汇报三个环节。中心岩溶景观资源专家陈伟海研究员介绍了广西自然景观资源概况及图集编制情况；随后，中心向广西自然资源厅赠送了《广西壮族自治区自然景观资源图集》；专题环节，来自中国、法国、英国的5位专家围绕岩溶景观资源的调查、开发与保护做了专题报告。

Prof. Chen Weihai from IRCK introduced the general situation of natural landscape resources in Guangxi
中心专家陈伟海研究员介绍广西自然景观资源概况

Chapter 4 International Exchange and Training

上左图 /Top left

Prof. Jiang Zhongcheng, the GB-II member presented the *Natural Landscape Resources Atlas of Guangxi Zhuang Autonomous Region* to Guangxi DNR

中心理事蒋忠诚研究员向广西自然资源厅赠送《广西壮族自治区自然景观资源图集》

上中图 /Top middle

Dr. Jean Bottazzi, a professional caver from France gave a presentation on "The exploration of the Asian longest cave, Shuanghe Cave in Suiyang"

法国探洞专家让·波塔次做题为"亚洲最长洞穴——绥阳双河洞勘探"的报告

上右图 /Top right

Prof. Zhang Yuanhai of IRCK, also the president of the Union of Asian Speleolgy gave a presentation on "The contributions of survey on karst geological heritages to the construction of nature reserves"

中心专家、亚洲洞穴联合会主席张远海研究员做题为"岩溶地质遗迹调查对自然保护地建设的贡献"的报告

下左图 /Bottom left

Dr. Philip John Rowsell, a professional caver from England gave a presentation on "The significant value of karst and caving landscape"

英国探洞专家菲利普·约翰·罗素做题为"岩溶与洞穴景观价值的重要意义"的报告

下中图 /Bottom middle

Dr. Shi Wenqiang from IRCK gave a presentation on "The survey on geological heritages and the geological cultural villages construction in southwest karst area, China"

中心专家史文强博士做题为"中国西南岩溶区地质遗迹调查与地质文化村建设"的报告

下右图 /Bottom right

Chen Lixin, a professional caver from China, gave a presentation on "Cave exploration and geopark"

中国探洞专家陈立新做题为"洞穴探险与地质公园"的报告

3. The Seminar on China-ASEAN Karst Geology Comparison Study and Mapping (2021)

2021 年中国 – 东盟岩溶地质对比研究与编图研讨会

IRCK hosted the Seminar on China-ASEAN Karst Geology Comparative Study and Mapping on 21 May 2021, during the China-ASEAN Mining Cooperation Forum, with more than 90 scholars from 14 countries participated both virtually and on-site. Presided over by Mr. Zhao Xiaoming, the deputy director of the Institute of Karst Geology, the seminar had Mr. Wu Xixi, the GB-II member of IRCK, also the party group member of the Department of Natural Resources of Guangxi, attend the event and deliver a welcome speech. China attaches high importance to geosciences cooperation with ASEAN and the partners along the Belt and Road, hoping that the seminar may promote cooperation with related countries to develop karst science jointly, and may enable the sustainability of resources and environment in karst areas over the world.

2021 年 5 月 21 日，中国 – 东盟矿业合作论坛召开期间，中心承办了中国 – 东盟岩溶地质对比研究与编图研讨会，来自 14 个国家的 90 多名专家学者通过"线下 + 线上"的方式共同参与了

左图 /Left
Mr. Zhao Xiaoming, the deputy director of IKG, presided over the seminar
中国地质科学院岩溶地质研究所副所长赵小明主持本次会议

右图 /Right
Mr. Wu Xixi, GB-II member of IRCK and the party group member of the Department of Natural Resources of Guangxi attended the meeting and delivered a speech
中心理事、广西壮族自治区自然资源厅党组成员吴锡熹致辞

本次会议。中国地质科学院岩溶地质研究所副所长赵小明主持会议，中心理事、广西壮族自治区自然资源厅党组成员吴锡熹先生出席会议并致辞。我国高度重视与东盟以及共建"一带一路"国家之间的地学合作，希望通过本次研讨会推动与相关国的合作，从而促进岩溶科学进步，推进全球岩溶区资源和环境的可持续性。

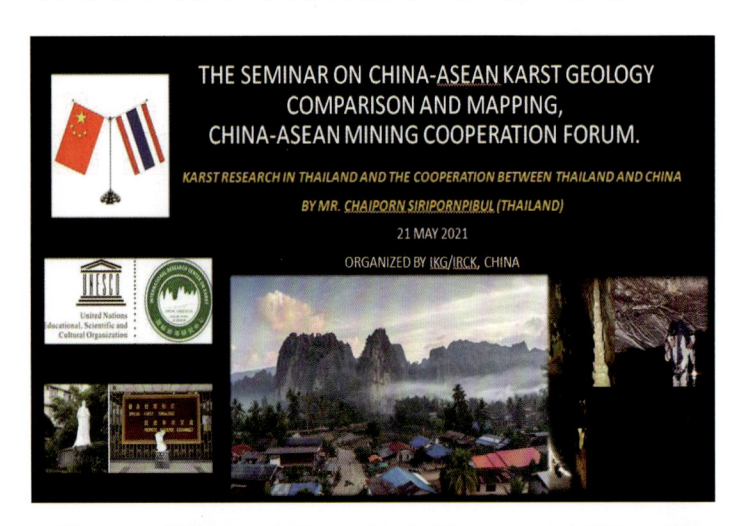

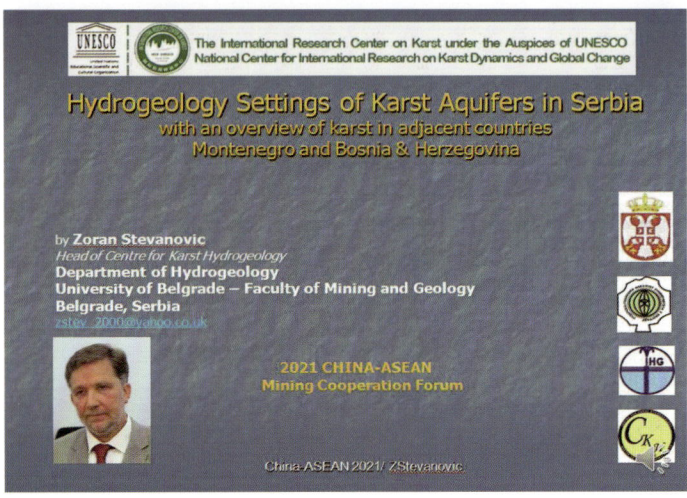

上图/Top

Mr. Chaiporn Siripornpibul (Thailand): Karst research in Thailand and the cooperation between Thailand and China
中心特邀专家柴鹏·斯里蓬皮布尔先生（泰国）做题为"泰国岩溶研究及中泰合作"的报告

下图/Bottom

Mr. Zoran Stevanovic (Serbia): Hydrogeology settings of karst aquifers in Serbia with an overview of karst in adjacent countries Montenegro and Bosnia & Herzegovina
特邀专家佐伦·史蒂夫诺维奇（塞尔维亚）做题为"塞尔维亚岩溶含水层水文地质背景及邻国黑山与波黑岩溶概况"的报告

第四章　国际交流与培训

上图/Top

Mr. Martin Knez (Slovenia): Classical Karst Kras—Its Principal Characteristics and Geology

特邀专家马丁·内兹（斯洛文尼亚）做题为"经典岩溶——其主要特征与地质背景特点"的报告

下图/Bottom

Mr. Zargham Mohammadi (Iran): The contribution of dye tracing tests in the characterization of karst aquifers in Iran

特邀专家扎格汉·莫罕穆迪（伊朗）做题为"示踪实验在研究伊朗岩溶含水层特征中的应用"的报告

Chapter 4 International Exchange and Training

上左图 /Top left
Mr. Lyu Yong (China): General karst development along the New Western Land-Sea Corridor
吕勇（中国）：西部陆海新通道岩溶发育概况

上右图 /Top right
Mr. Kang Zhiqiang (China): The primary plan about China-Cambodia joint science and technology research on karst critical zone
康志强（中国）：中柬岩溶关键带科学与技术联合研究中心平台建设设想

下左图 /Bottom left
Mr. Zhang Cheng (China): The construction of the China-Slovenia "The Belt and Road" Joint Laboratory on Karst Geology
章程（中国）：中国-斯洛文尼亚岩溶地质"一带一路"联合实验室

下右图 /Bottom right
Mr. Xu Qi (China): China-Southeast Asia joint compilation of serial maps on karst geology and the environment
许琦（中国）：中国与东南亚地区岩溶地质环境合作编图

4.1.1.4 The Parallel Forums on Sustainable Utilization of Landscape Resources of China-ASEAN International Forum on Sustainable Development and Innovative Cooperation
中国－东盟可持续发展创新合作国际论坛景观资源可持续发展系列论

Since 2018, the year that Guilin was listed as a National Innovative Demonstration Area for the Sustainable Development Agenda, IRCK has shared its experience positively with Guilin, and organized three international forums on sustainable utilization of karst landscape resources with the Guilin Government jointly, proposing suggestions for local sustainable development proactively.

自 2018 年桂林被列为国家可持续发展议程创新示范区以来，中心积极分享岩溶资源可持续开发与利用的经验，与桂林市政府共同举办了三届岩溶景观资源可持续利用国际论坛，为地方可持续发展建言献策。

1. The Parallel Forum on Sustainable Use and Development of Landscape Resources of China-ASEAN International Forum on Sustainable Development and Innovative Cooperation & China (Guilin) High-level International Forum on Health Tourism (2018)

中国－东盟可持续发展创新合作国际论坛暨中国（桂林）国际健康旅游高端论坛景观资源可持续利用平行论坛（2018）

On 22 November 2018, the China-ASEAN International Forum on Sustainable Development and Innovative Cooperation & China (Guilin) High-level International Forum on Health Tourism was grandly opened in Guilin, Guangxi. About 400 domestic and foreign experts and scholars as well as enterprises representatives from 24 countries and international organizations including UNDP in China, the Asian Development Bank as well as Cambodia, Denmark, the Republic of Korea, Thailand, and the US, attended the opening ceremony to discuss about the sustainable use of landscape resources and the integrated development of health and tourism. In the afternoon, IRCK organized the Parallel Forum on Sustainable Utilization of Landscape Resources with the People's

Municipal Government of Guilin. The forum invited Mr. Bai Songtao, deputy secretary of the CPC Committee in Guilin, to deliver a welcome speech; and the experts from Serbia, Austria, the US, Denmark, and China to give keynote speeches. Mr. Yuan Daoxian, the GB-II member of IRCK, also the academician of the Chinese Academy of Sciences, and Mr. Wang Yanxin, the GB-II member of IRCK, also the president of China University of Geosciences (Wuhan), attended the forum and discussed with the presenters. On the following day, the attendees visited the Geological Museum of Guilin University of Technology and Huixian Karst Wetland Park, acquired better understanding about the environmental changes of the wetland and its economic benefits to locals. The forum was encouraging the cooperation to promote the sustainable development of different countries.

2018年11月22日，"中国－东盟可持续发展创新合作国际论坛暨中国(桂林)国际健康旅游高端论坛"在广西桂林隆重开幕。联合国开发计划署驻华代表处、亚洲开发银行，以及柬埔寨、丹麦、韩国、泰国、美国等24个国家和国际组织的约400名国内外专家学者、企业代表等参会，围绕景观资源可持续利用、健康旅游融合发展等领域开展深入探讨。当日下午，中心联合桂林市人民政府组织了中国－东盟可持续发展论坛景观资源可持续利用平行论坛。论坛邀请了桂林市委副书记白松涛做主旨发言，同时邀请了塞尔维亚、奥地利、美国、丹麦和中国的专家做主旨报告。中心理事袁道先院士，中心理事、中国地质大学（武汉）王焰新校长等参与了交流研讨。11月23日，中心组织与会专家参观了桂林理工大学地质博物馆和会仙湿地公园，为与会者介绍了湿地公园生态环境变化，以及为周边居民带来的可持续经济效益，鼓励大家开展合作，共同推进各国可持续发展。

上左图 /Top left
Mr. Zoran Stevanovic (Serbia): Karst aquifers of SE Europe and East Mediterranean Basin—One of the World's most richest treasuries of groundwater
佐伦·史蒂夫诺维奇（塞尔维亚）：世界最丰富地下水资源之一——欧洲东南部和地中海盆地东部岩溶含水层研究进展

下左图 /Bottom left
Mr. Ralf Benischke (Austria): Karst groundwater monitoring, targets, strategies, and evaluation
拉尔夫·比尼斯尔克（奥地利）：岩溶地下水监测、目标、策略与评估

上右图 /Top right
Mr. Petar Milanovic (Serbia): Karst aquifers and water resources development of Bosnia and Herzegovina
皮特·米拉诺维奇（塞尔维亚）：波黑岩溶含水层与水资源开发

下右图 /Bottom right
Mr. Jiang Zhongcheng (China): Characteristics, formation cause and resources potentials of karst landform in Guilin
蒋忠诚（中国）：桂林岩溶景观特征、成因及资源潜力

Chapter 4　International Exchange and Training

上左图 /Top left
Mr. Stefan Werner (Denmark): Construction experience of urban green and blue infrastructures in Germany and Sweden
史蒂凡·维尔纳（丹麦）：德国与瑞典城市绿色和蓝色基础设施建设经验

下左图 /Bottom left
Mr. Cao Jianhua (China): Progress of International Big Scientific Plan on "Global Karst"
曹建华（中国）："全球岩溶"国际大科学计划进展

上右图 /Top right
Mr. George Veni (the USA): Major concepts in cave and karst management strategy
乔治·维纳（美国）：洞穴与岩溶管理策略

下右图 /Bottom right
Scientists from IRCK introduced the Huixian Wetland Park to the attendees
中心科技人员向参会嘉宾介绍会仙湿地公园

The Second 6 Years of IRCK

2. China—ASEAN International Forum on Sustainable Development and Innovation Cooperation-The Parallel Forum on the Sustainable Utilization of Karst Landscape Resources (2019)

中国-东盟可持续发展创新合作国际论坛——喀斯特景观资源可持续利用平行论坛（2019）

On September 19, 2019, IRCK held its Parallel Forum on Sustainable Utilization of Karst Landscape Resources during the China-ASEAN International Forum on Sustainable Development and Innovative Cooperation, with more than 80 experts and scholars from 17 countries like Brazil, Cambodia, France, and China attended. The forum hoped to promote the sustainable use and development of landscape resources in Guilin and experiences after exchange and communication. The topics include karst landscapes sustainable utilization, the reasonable development of water resources, as well as the functions for karst carbon cycle in climate change.

左图 /Left
Prof. Harrison Pienaar (South Africa): Sustainable water resources planning and management—A case study of water resources utilization in KwaZulu-Natal province, South Africa
哈里森·皮纳尔（南非）：可持续水资源规划和管理——以南非夸祖鲁-纳塔尔省水资源利用为例

右图 /Right
Prof. Ralf Benischke (Austria): Quality and management of karst water in Austria
拉尔夫·比尼斯尔克（奥地利）：奥地利岩溶水质量与管理

2019年9月19日，在中国-东盟可持续发展创新合作国际论坛召开期间，中心主办了喀斯特景观资源可持续利用平行论坛，来自巴西、柬埔寨、法国和中国等17个国家的专家、学者共计80余人参加了此次论坛。本次平行论坛旨在继续推动桂林景观资源可持续利用，通过合作交流，获取宝贵经验。论坛围绕岩溶景观资源、水资源的合理开发利用、岩溶碳循环在气候变化过程中的作用等进行了深入交流。

上图 /Top

Prof. Chen Weihai (China): Ideas on the key technology research, development, and demonstration of karst landscapes sustainable utilization in Lijiang River Basin

陈伟海（中国）：漓江流域岩溶景观可持续利用关键技术研发与示范

下图 /Bottom

Prof. Chaiporn Siripornpibul (Thailand): Status of CK management in Thailand after cave rescue

柴鹏·斯里蓬皮布尔（泰国）：Tham Luang 洞穴救援后的泰国洞穴与岩溶综合管理现状

第四章 国际交流与培训　217

上图/Top
Prof. Sasa Milanovic (Serbia): Problems and solutions of water shortage on the different hydrogeological conditions—Example from north Somalia (Africa) and Dinaric karst (Europe)
萨沙·米拉诺维奇（塞尔维亚）：不同水文地质条件下水资源短缺的问题与解决途径——以北索马里（非洲）和第纳尔岩溶区（欧洲）为例

下图/Bottom
Prof. Cao Jianhua (China): New approach for carbon sequestration Karst carbon cycle and carbon sink effect
曹建华（中国）：碳汇新途径——岩溶碳循环与碳汇效应

国际岩溶研究中心第二个六年历程

3. The Seminar of "Guilin Practice: Sustainable Utilization of Typical Karst Landscape Resources and Coordinated Development of Ecological Industry" under the China-ASEAN International Forum on Sustainable Development and Innovation Cooperation (2020)

中国－东盟可持续发展创新合作国际论坛主题研讨会"典型喀斯特景观资源可持续利用及生态产业协同发展的桂林实践"（2020）

On 28-29 November 2020, at the China-ASEAN International Forum on Sustainable Development and Innovation Cooperation hosted by the Ministry of Science and Technology and the People's Government of Guangxi Zhuang Autonomous Region, organized by China-ASEAN Technology Transfer Center of Guangxi Zhuang Autonomous Region, Department of Science and Technology of Guangxi, and Guilin Municipal People's Government, IRCK organized a seminar on "Guilin Practice: Sustainable Utilization of Typical Karst Landscape Resources and Coordinated Development of Ecological Industry". The seminar is an important exchanging platform for continuous promotion of the coordinated development of karst landscape resources and ecological industries under the epidemic, hoping to promote the construction of Guilin as a National Innovative Demonstration Area for the Sustainable Development Agenda.

2020年11月28~29日，在科学技术部和广西壮族自治区人民政府主办，中国－东盟技术转移中心、广西壮族自治区科学技术厅和桂林市人民政府承办的"中国－东盟可持续发展创新合作国际论坛"上，中心组织召开了"典型喀斯特景观资源可持续利用及生态产业协同发展的桂林实践"主题研讨会。本次会议是疫情下继续推进岩溶景观资源与生态产业协同发展的重要交流平台，旨在持续推进桂林国家可持续发展议程创新示范区建设。

At the meeting, the project leader introduced the development and demonstration of the key technologies for sustainable utilization of karst landscape resources in the Lijiang River Basin following the aspects, the ecological industrialization and the sustainable utilization of water resources of Guilin landscape. The experts, project backbones, representatives, counties, districts, and enterprises of Guilin had a thorough discussion on the issues like current status of karst landscape resources, sustainable development, bottlenecks for ecological industry development, difficulties in water resources regulation and utilization, big data services and other technical support, decision-making and government management, etc.

会上，项目负责人介绍了漓江流域喀斯特景观资源可持续利用关键技术研发与示范、漓江流域景观资源生态产业化关键技术研究与示范、桂林景观水资源可持续开发利用关键技术研发与示范的进展情况。特邀专家、项目骨干、各市县区代表及企业代表围绕喀斯特景观资源现状和可持续发展问题、生态产业发展瓶颈、水资源调控利用难点、技术支撑与大数据服务、决策与政府管理等多个方面展开了激烈探讨。

The attendees at the seminar
主题研讨会现场

4.1.1.5 Serial China-Africa Water Resources Dialogues
中非水资源论坛系列会议

China-Africa Water Resources Dialogue is a serial forum dedicated to promoting the sustainable use of water resources in developing countries in China and Africa, which is initiated by Prof. Xu Yongxin, the AC-II member and professor of the University of the Western Cape in South Africa. From 2016 to 2021, as one of the sponsors, IRCK organized or co-organized three forums held in 2016, 2017, and 2018 respectively, hoping to provide a platform for efficient and systematic sharing of theoretical and practical experience in the utilization of water resources in China and Africa.

中非水资源论坛是由中心学术委员、南非西开普大学教授徐永新发起的致力于推进中国和非洲地区发展中国家水资源可持续利用的系列论坛。2016~2021年，中心作为主要发起单位之一参与了2016年、2017年、2018年三届论坛的组织工作，旨在为中国和非洲地区的水资源利用提供高效、系统的理论与实践经验的分享舞台。

1. The 4th China-Africa Water Forum (2016)
第四届中非水资源论坛（2016）

Sponsored jointly by IRCK and the University of the Western Cape in South Africa, the International Symposium on Sustainable Utilization of Water Resources in Developing Countries (the 4th China-Africa Water Forum) took place in Taiyuan of China, Aug. 1–3, 2016. The Taiyuan University of Technology and the Institute of Karst Geology were the organizers.

由中心和南非西开普大学共同主办，太原理工大学与中国地质科学院岩溶地质研究所共同承办的发展中国家水资源可持续利用国际研讨会暨第四届中非水资源论坛于2016年8月1~3日在山西太原成功举办。

Mr. Yuan Daoxian, the GB-II member and academician of the Chinese Academy of Sciences, delivered a keynote speech on socio-economic development and current resources and environmental status in China, highlighting 16 major challenges on karstology that need to be addressed. Then, Prof. Cao Jianhua, the executive deputy director of IRCK presented the preparation of the Big Scientific Plan on "Global Karst", which attracted strong interest from all the participants.

中心理事、学术委员会主任袁道先院士就我国社会经济发展及资源环境现状做了精彩的主旨发言，提出了 16 个亟待解决的岩溶科学问题；中心常务副主任曹建华研究员向各国参会代表介绍了中心正在发起的"全球岩溶动力系统资源环境效应"国际大科学计划的编写进展，获得了国内外与会人员的高度关注。

A total of 10 prestigious scientists on water science delivered keynote speeches, while 32 young and middle-aged scholars presented their latest findings on water resources exploitation and utilization. The 3-day event focused on many issues related to water resources and environment: the impact of mining on water environment; the impact of climate change on regional floods and droughts; as well as water resources cycle, exploitation, management, and sustainable utilization. The event facilitated sharing, cooperation, communication, and friendship between China and Africa.

10 位水科学领域专家围绕大会主题发表了精彩的主旨演讲，32 位中青年专家学者交流了水资源开发利用的最新研究成果。为期三天的中非水资源论坛围绕矿产开采对水生态环境影响，气候变化对区域旱涝灾害的影响，水资源循环、开发、管理和可持续利用等进行了深入研讨，促进了水资源可持续利用经验分享，促进了中非水科学领域的交流与合作，增进了中非科技人员之间的友谊。

More than 150 scientists from ten countries (including China, South Africa, Ghana, Zimbabwe and India) participated. Mr. Liu Tongliang, director of IRCK, Dr. Liang Liping, vice president of Taiyuan University of Technology, Mr. Wang Hao, the academician of Chinese Academy of Engineering and Mr. Matthys A. Dippenaar, the chair of South Africa Chapter of IAH and Mr. Xu Yongxin, AC-II member of IRCK attended the opening ceremony.

中心主任刘同良、太原理工大学副校长梁丽萍、中国工程院院士王浩、国际水文地质学家协会南非委员会主席马斐斯·第皮纳尔、中心学术委员徐永新教授等出席了开幕式。来自中国、南非、加纳、津巴布韦、印度等 10 个国家和地区的 150 余位科技人员参加了此次会议。

Chapter 4   International Exchange and Training

The opening ceremony of the forum
会议开幕式现场

2. The 5th China-Africa Water Resources Dialogue (2017)
第五届中非水资源论坛（2017）

On July 29 to August 6, 2017, the 5th China-Africa Water Resources Dialogue was held in Zimbabwe, with the National University of Science and Technology of Zimbabwe (NUST) as the organizer and IRCK as the co-organizer. The dialogue set 8 topics: the management and sustainable utilization of water resources, the impact of mining on water resources and ecosystems, the impact of coalbed methane and shale gas on groundwater, exploration and development of groundwater resources, transboundary water issues and policies, promotion of the relationship between China and Africa, water recycling and transformation of "four types of water" (atmospheric water, surface water, soil water, and groundwater), and the impact of climate change on regional droughts and floods. Besides, the IRCK delegation discussed the establishment of monitoring stations in Zimbabwe with the dean, the vice dean of NUST, and Prof. Xu Yongxin (the AC-II member), hoping to promote the Big Scientific Plan on "Global Karst" jointly.

2017年7月29日~8月6日，第五届中非水资源论坛在津巴布韦举办，本次论坛由津巴布韦国立科技大学主办，中心协办。会议围绕水资源管理与可持续利用、采矿对水资源与生态系统的影响、煤层气和页岩气对地下水的影响、地下水资源的勘探和开发、跨境水资源问题和政策、促进中国和非洲之间的密切合作、水资源循环和"四水"转化、气候变化对区域干旱和洪涝灾害的影响等8个专题展开。会后，中心代表团与津巴布韦国立科技大学校长、副校长、中心学术委员徐永新教授共同讨论建立观测站事宜，合作推进"全球岩溶"国际大科学计划。

Chapter 4 International Exchange and Training

IRCK delegation discussed with the dean and his colleagues of NUST about the establishment of the monitoring stations
中心代表团与津巴布韦国立科技大学校长一行商讨建站等合作事宜

3. The 6th China-Africa Water Resources Dialogue (2018)
第六届中非水资源论坛（2018）

On July 21-28, 2018, the 6th China-Africa Water Resources Dialogue was held in Egypt, with IRCK as the co-organizer. The dialogue set 4 topics: the impact of mining on aquatic ecological environment, the impact of climate change on regional droughts and floods, as well as the cycles, development, management, and sustainable utilization of water resources. Taking the International Big Scientific Plan on "Global Karst" as a great opportunity for cooperation, IRCK delegation had extensive exchanges with other attendees, looking forward to a wonderful collaboration between China and Africa through projects and mutual visits.

2018年7月21~28日，第六届中非水资源论坛在埃及举办，中心为本次会议协办方。会议围绕矿产开采对水生态环境影响，气候变化对区域旱涝灾害的影响，水资源循环与开发、管理和可持续利用等展开研讨。此次会议，中心以推进"全球岩溶"国际大科学计划为契机，与参会代表展开了广泛的交流沟通，期待通过项目合作、人员互访等推进中非岩溶科学合作。

Group photo of IRCK delegates with the attendees
中心代表团与会议嘉宾合影

4.1.1.6 Serial Symposiums on Karst of Xiangxi Geopark
湘西喀斯特国际学术研讨会

As the technical support organization for the application of Xiangxi UNESCO Global Geopark (UGGP), IRCK hosted two sessions of international symposiums with the People's Government of Xiangxi Prefecture in 2017 and 2018 respectively, aiming to advertise the karst geological heritage of Xiangxi and enhance its popularity. Meanwhile, it also hopes to provide good experience and advanced technologies for the development, utilization and protection of geological heritage in Xiangxi.

中心作为申报湘西世界地质公园的技术支撑单位，分别于 2017 年、2018 年与湘西州人民政府共同主办了两次国际学术研讨会，旨在宣传湘西岩溶地质遗迹，提升其知名度，同时也为湘西地质遗迹的开发、利用与保护提供成熟经验与先进技术。

1. The International Symposium on Karst of Xiangxi Geopark (2017)
2017 年湘西喀斯特国际学术研讨会 (2017)

On July 1, 2017, the first International Symposium on Karst of Xiangxi Geopark was held in Jishou, with the People's Government of Xiangxi Prefecture and IRCK as its organizers. The symposium has 40 attendees including Academician Yuan Daoxian (the GB-II member), Prof. Jiang Zhongcheng (the GB-II member), Prof. John Gunn from the University of Birmingham in UK, Prof. Andrej Kranj from Slovenian Karst Research Institute, Prof. Julia James from the University of Sydney, also the vice chairman of the Great Blue Mountains World Natural Heritage Committee, Prof. Alexander Klimchouk from the Institute of Geological Sciences of the National Academy of Sciences of Ukraine, and other experts and scholars from Central South University, China University of Geosciences (Wuhan), and Southwest University, among others.

2017 年 7 月 1 日，由湘西州政府、国际岩溶中心联合主办的湘西喀斯特国际学术研讨会在吉首召开。中心理事袁道先院士、中心理事蒋忠诚研究员，英国伯明翰大学约翰·甘恩教授，斯洛文尼亚岩溶地质研究所安德烈·克朗捷教授，澳大利亚悉尼大学教授、大蓝山世界自然遗产委员会副主席茱莉亚·詹姆斯，乌克兰国家科学院地质科学研究所亚历山大·克利姆乔教授，以及来自中南大学、中国地质大学（武汉）、西南大学等高校和科研院所的专家学者共40余人参加了研讨会。

  After comparing the geological heritage in Xiangxi with that of Guilin, Academician Yuan stated that applying for UGGp requires taking local advantages, especially the ones with global significance, exploring the uniqueness worldwide, and great advertisement; in addition, he mentioned that it was also necessary to strengthen scientific research, find out the formation and characteristics of geological heritage, with more papers published in important international academic journals timely; moreover, he thought that it should make more efforts for science popularization to sustain tourism development, and decision-makers should put related measures for protection into effect. He emphasized that Xiangxi is possible to become a tourism resort integrating leisure and popular science based on its richness in natural resources (e.g. geological resources) and folk cultures, with the common efforts of administrators, research institutions, and local communities. Prof. Jiang concluded that all the attended experts agreed that there are a large number of world-class geological heritage and landscapes in Xiangxi, with instructive suggestions like more efforts should be made in scientific investigation and research, which could be considered as the "guide" for the future work, he emphasized that IRCK will contribute full support to the application of Xiangxi UGGp (Quoted from the news on official website of Department of Natural Resources of Hunan Province).

  袁道先院士将湘西的岩溶地质遗迹与桂林的岩溶地质遗迹进行对比后指出，申报世界地质公园要抓住当地具有世界级特色优势的资源，探究地质遗迹的独特性，并推介给全世界；要加强科学研究，探明地质遗迹成因与特点，在国际重要学术期刊上及时发表科研成果；要加大科普力度，从地质旅游方面推动旅游产业可持续发展；要加强管理，落实地质遗迹保护措施。他强调，湘西地质遗迹丰富多彩，在管理者、科研机构、群众等各方努力下，湘西可打造成以地质资源等自然生态和民俗文化为主的休闲胜地、科普基地、旅游目的地。蒋忠诚理事指出参会专家一致认为湘西拥有大量的世界级地质遗迹和景观，对湘西申报世界地质公园提出了进一步加强科考科研等建议意见，为湘西申报世界地质公园的下一步工作提供了"指南"，并强调中心将全力支持湘西申报世界地质公园（摘自湖南省自然资源厅官网）。

Academician Yuan gave a keynote speech
袁道先院士做主旨发言

2. The 2nd International Symposium on Karst of Xiangxi Geopark (2018)
湘西地质公园第二届喀斯特国际学术研讨会（2018）

On Nov. 17-18, 2018, the 2nd International Symposium on Karst of Xiangxi Geopark was held in Xiangxi Prefecture, with IRCK and the People's Government of Xiangxi Tujia & Miao Autonomous Prefecture as organizers, the Bureau of Land and Resources of Xiangxi Tujia & Miao Autonomous Prefecture, the Tourism and Overseas Chinese and Foreign Affairs Commission of Xiangxi Tujia & Miao Autonomous Prefecture, and the Administration of Xiangxi Geopark as the co-organizers. The event was attended by 70 experts and scholars from 19 countries.

2018年11月17~18日，中心协同湘西自治州人民政府主办，湘西自治州国土资源局、湘西自治州旅游和外事侨务委员会、湘西地质公园管理处承办的湘西地质公园第二届喀斯特国际学术研讨会在湘西自治州召开，来自19个国家的70名专家学者参加了会议。

The symposium lasted for two days and consisted of two parts: a field study and an academic workshop. On Nov.17, the attendees visited karst landscapes (i.e. Dehang Grand Canyon, Sisters Peak Rock Pillar Group, GSSP at Guzhangian, Guzhang Red Stone Forest, and Dalong Cave Waterfall), and the historic and cultural landscapes (i.e. Aizhai Bridge and Furong Ancient Town). On Nov 18, domestic and foreign experts and scholars had a thorough discussion on Xiangxi aspiring UGGp.

本次学术研讨会为期2天，分为野外考察和学术研讨两个部分。11月17日，专家先后考察了湘西地质公园内的德夯大峡谷、姊妹峰岩柱群、古丈阶"金钉子"、古丈红石林、大龙洞瀑布等岩溶地质景观和矮寨大桥、芙蓉古镇等人文历史景观。11月18日，国内外专家就湘西地质公园申报世界地质公园进行了充分交流。

Chapter 4 International Exchange and Training

上左图 /Top left
Comments by Prof. Petar Milanovic (Serbia), the GB-II member of IRCK
中心理事皮特·米拉诺维奇先生（塞尔维亚）在会上发言

下左图 /Bottom left
Keynote speech by Ms. Aleksandra Maran Stevanovic, the special invited guest and advisor to the Natural History Museum of Serbia
中心特邀嘉宾、塞尔维亚自然历史博物馆科学顾问亚历桑德拉·马兰·史蒂夫诺维奇女士做主旨发言

上右图 /Top right
Keynote speech by Prof. George Veni (the USA), the AC-II member and the president of UIS
中心学术委员、国际洞穴联合会主席乔治·维纳（美国）做主旨发言

下右图 /Bottom right
The participants were visiting Guzhang Red Stone Forest and discussing its formation
中外嘉宾在古丈红石林景区探讨红石林成因

The Second *6* Years of IRCK

4.1.1.7 Serial forums of Chinese Karst Experts
中国岩溶专家系列论坛

Karst in China is widely distributed and has obvious characteristics. China's karst resources and environmental problems are prominent, with complex causes and various solutions. In order to integrate the scientific research achievements of Chinese karst experts and create a Chinese mode for the world to solve karst resources and environmental problems, IRCK held two sessions of "Chinese Karst Experts Forum" under the initiative of Academician Liu Congqiang and Academician Yuan Daoxian.

中国的岩溶分布面积广泛，特色明显。中国的岩溶资源环境问题突出，成因复杂，解决途径多样。为集成中国岩溶专家的科研成果，为世界打造解决岩溶资源环境问题的中国模式，中心在中国科学院院士刘丛强先生及中心理事袁道先院士的倡议下召开了两届"中国岩溶专家论坛"。

1. The 1st Chinese Karst Experts Forum (2017)
首届中国岩溶专家论坛（2017）

On October 23, 2017, IRCK held the 1st Chinese Karst Experts Forum in Guilin, aiming to set up an exchange platform for domestic karst research, and critical for karst research in China. Academician Liu Congqiang, then vice president of National Natural Science Foundation of China (NSFC), proposed an initiative to establish "Karst System Science Partnership" (KSSP) firstly. He made a preliminary plan for its objectives, tasks, and scientific positioning of KSSP, expressing his expectation to improve the research level of karstology in China and enhance its international influence. His speech was followed by Mr. Li Weihong, then deputy inspector of the Department of Science and Technology of Guangxi (DSTG), who spoke highly of the great contributions that IRCK has made to promote local ecological and economic development over the recent years. He also indicated that he will continue to support the work of IRCK and other karst related research institutions to make joint contribution to national karst research with all the experts in China, and to work closely with other domestic experts to further advance the development of karstology in China.

2017年10月23日，由中心主办的第一届"中国岩溶专家论坛"在桂林开幕，此次会议旨在搭建国内岩溶研究交流平台，凝聚国内岩溶研究核心力量。中心特邀专家、国家自然科学基金委员会副主任刘丛强院士提出了建立"岩溶系统科学联盟"（KSSP）的倡议，并对其任务目标、科学定位等进行了初步规划，表达了对提高国内岩溶研究水平、提升国际影响力的希望和愿景。广西科技厅副巡视员黎卫红先生代表广西科技厅就中心近年来为地方生态经济建设做出的贡献进行了高度评价，表达了对中心及其他相关单位的一贯支持，并表明未来将与国内其他专家一同努力，为国家岩溶事业做出贡献。

A total of 12 experts include Academician Yuan Daoxian (GB-II member) and Prof. Jiang Zhongcheng (GB-II member), Prof. Cheng Hai (AC-II member) and Prof. Jiang Yongjun (AC-II member), with specially invited professors, Prof. Wang Kelin, Prof. Wang Yu, Prof. Li Xiankun, Prof. Hu Xiaonong, Prof. Chen Xi, Prof. Gao Xubo, Prof. Pang Zhonghe and Prof. Cao Jianhua (the executive deputy director) made wonderful reports on karst in China, covering the overall features of karst in China, karst water resources, geothermal resources, rocky desertification control, climate change, etc. These reports are strong supporting for the solutions of karst resources and environmental problems.

中心理事袁道先院士、蒋忠诚研究员，中心学术委员程海教授、蒋勇军教授，以及中心特邀专家王克林、王宇、李先琨、胡晓农、陈喜、高旭波、庞忠和教授及中心常务副主任曹建华研究员等12位专家做了精彩报告。报告分别从中国岩溶总体特色到岩溶水资源、地热资源、石漠化治理、气候变化等系统介绍了中国岩溶的研究进展，为后期解决岩溶资源环境问题提供了强劲支撑。

The group photo for the first forum
参会人员合影

2. The 2nd Chinese Karst Experts Forum and the Seminar on the Comparison and Mapping of Karst Geological Environment in Key Areas along the Belt and Road (2020)

第二届中国岩溶专家论坛暨"一带一路"重点区岩溶地质环境对比与编图交流会（2020）

On 19 November 2020, the Second Chinese Karst Experts Forum and the Seminar on the Comparison and Mapping of Karst Geological Environment in Key Areas along the Belt and Road was held in Guilin for better exchange academic ideas and broaden the international vision. The event was organized by IKG/IRCK, and co-organized by National Innovation Alliance of Southwest Karst Desertification Control under National Forestry and Grassland Administration, with Prof. Cai Yunlong from Peking University as the chairperson.

2020年11月19日，第二届中国岩溶专家论坛暨"一带一路"重点区岩溶地质环境对比与编图交流会在桂林召开，本次会议旨在开展"一带一路"岩溶地质领域学术交流，拓宽国际视野。会议由中国地质科学院岩溶地质研究所/联合国教科文组织国际岩溶研究中心主办，国家林业和草原局西南岩溶石漠化治理国家创新联盟协办。北京大学蔡运龙教授主持了本次会议。

In the welcome speech by Prof. Jiang Zhongcheng, he stated the importance of karst research along the Belt and Road, analyzed karst research advantages in China than other part of the world, and highly expected more fruitful cooperation results with the countries along the Belt and Road. All the attendees had in-depth communications and discussions on karst geology along the Belt and Road. More than 30 experts from institutions like Peking University, National Forestry and Grassland Administration, the Chinese Academy of Forestry, Beijing Forestry University, Xishuangbanna Tropical Botanical Garden of Chinese Academy of Sciences, Institute of Hydrobiology of Chinese Academy of Sciences, Guangxi Institute of Botany of Chinese Academy of Sciences, and Yunnan Geological Survey participated.

中心理事蒋忠诚研究员强调了以我国为主开展"一带一路"沿线岩溶地质调查研究的重要性，分析了我国岩溶地质研究在世界范围内的优势，并对未来开展"一带一路"沿线岩溶地质国际合作寄予期望。来自全国各地的岩溶学者围绕"一带一路"沿线岩溶地质特色展开了深入交流与讨论。本次会议共有来自北京大学、国家林业和草原局、中国林业科学研究院、北京林业大学、中国科学院西双版纳热带植物园、中国科学院水生生物研究所、中国科学院广西植物研究所、云南省地质调查局等单位的30余名专家代表出席。

第四章　国际交流与培训

上图 /Top
Group photo of the experts
参会人员合影

下图 /Bottom
The experts at the meeting
会议交流现场

国际岩溶研究中心第二个六年历程

4.1.1.8 Other important international or domestic academic conferences organized by IRCK
其他重要国际国内学术会议

1. Post-congress tour on karst in southwest China after the 33rd IGC
第 33 届国际地理大会会后考察

The 33rd International Geographical Congress (IGC) took place in Beijing from 21st to 25th August, 2016. IRCK organized the post-congress tour on karst with Yunnan University/International Joint Research Center on Karst under the Department of Science and Technology of Yunnan Province jointly, attracting 11 foreign scholars from 6 countries like the US, Australia, Poland, Russia, Italy, and Spain. Through discussions, Maocun Experimental Site excursion, the core zone of Guilin World Natural Heritage Site excursion, together with the Shilin and Jiuxiang excursion in Yunnan, IRCK popularized its achievements and displayed Chinese fantastic karst landscapes to the counterparts over the world.

第 33 届国际地理大会于 2016 年 8 月 21~25 日在北京顺利召开。25 日会议闭幕后，中心与云南大学 / 云南科技厅国际喀斯特联合研究中心共同承担岩溶专线会后考察，共有来自美国、澳大利亚、波兰、俄罗斯、意大利、西班牙 6 个国家的 11 名外籍学者参加。通过会谈研讨、桂林毛村野外试验场参观、桂林漓江流域喀斯特自然遗产地核心区考察、云南石林和云南九乡岩溶景观考察等，宣传中心成果，向世界同行展示中国特色岩溶地貌。

左图 /Left
The experts visited IKG/IRCK
专家团访问岩溶所 / 国际岩溶研究中心

右图 /Right
Prof. Liu Hong from Yunnan University led a group of experts to Shilin (Stone Forest)
云南大学刘宏教授带领专家团考察石林地貌

2. The International Symposium on Karst Critical Zones
岩溶关键带国际学术研讨会

On 19 July 2017, IRCK hosted the International Symposium on Karst Critical Zones in Kunming, with more than 50 delegates from 17 countries like the US, Brazil, Thailand, South Africa, and China attended. Through wonderful presentations and in-depth discussions, the attendees obtained a better understanding on the research progress of karst critical zones, geochemical models of karst systems, observation of critical zones, and hydrological processes.

2017年7月19日,中心在云南昆明主持召开了"岩溶关键带国际学术研讨会",来自美国、巴西、泰国、南非和中国等17个国家的50余名代表参加了会议。会议围绕岩溶关键带的研究进展、岩溶系统地球化学模型、关键带监测研究、水文过程等展开了精彩交流。

The International Symposium on Karst Critical Zones
岩溶关键带国际学术研讨会现场

从左至右、从上至下依次为

From left to right, top to bottom are

Cao Jianhua (China): Initial International Big Scientific Plan on "Resources and Environmental Effects of Global Karst Dynamic Systems"; Jonathan Arthur (the USA): A review of the hydrogeology and ecosystem restoration plans for the Edwards aquifer, Texas; Chen Zhu (the USA): Development of international critical zone observatory programs and geochemical modeling of reactions in karst systems; Chen Xi (China): Hydrological cycle in the karst critical zone—An integrated approach; Mitja Prelovsek (Slovenia): Toward a comprehensive quantitative model of karst CO_2 dynamic—A key drive in karst critical zone; Saša Milanović (Serbia): Hydrogeological engineering of karst aquifer—Problems and solutions; A. A. El-Fiky (Egypt): Hydrogeochemical evolution of Na-Cl karst spring waters of Siwa Oasis, Western Desert, Egypt; Zhen Hongbo (China): Birth of the Yangtze River — Tectonic-geomorphic implications; Mehran Maghsoudi (Iran): Karst landforms in Iran

曹建华（中国）："全球岩溶动力系统资源环境效应"国际大科学计划介绍；乔纳森·亚瑟（美国）：得克萨斯州爱德华含水层水文地质与生态系统恢复计划；Chen Zhu（美国）：国际关键带观测项目进展及岩溶系统系列反应的地球化学模型；陈喜（中国）：岩溶关键带水文循环——一种综合性研究方法；米提亚·普利罗斯克（斯洛文尼亚）：岩溶CO_2动力系统的综合定量研究模型；萨沙·米拉诺维奇（塞尔维亚）：岩溶含水层水文地质工程——问题及解决方案；艾尔·菲克（埃及）：埃及西部沙漠锡瓦绿洲（Siwa Oasis）岩溶泉Na-Cl水文地球化学进展；郑洪波（中国）：长江的形成与演化——构造、地貌的启示；梅鹤朗·玛撒迪（伊朗）：伊朗岩溶地貌

3. The Karst Ecology Session of the 12th International Congress of Ecology in Beijing

组织召开第 12 届国际生态学大会岩溶生态分会场

The 12th International Congress of Ecology (INTECOL2017), sponsored by the Ecological Society of China (ESC), convened at China National Convention Center on August 21, 2017. During the meeting, IRCK organized a session on karst ecology titled as "Karst ecosystem: significance, degradation, and restoration from local to global scales", which was attended by more than 60 domestic and foreign delegates. Eight scholars from Lakehead University in Canada, South China Botanical Garden of the Chinese Academy of Sciences, Beijing Forestry University, and other institutions made presentations including the comprehensive control mode of rocky desertification in karst areas, the evaluation of karst ecosystem functions, the interpretation of remote sensing images of rocky desertification areas in the graben basins of Southwest China, karst ecology and tourism, karst vegetation root system and ecological water use, the lime soil characteristics of karst ecosystem, the water and soil conservation mechanism of karst peak cluster-depression, and the response of soil nutrients to vegetation restoration in degraded karst areas. This session manifested an increasing attention on karst ecosystem from the ecology society, and IRCK will act as a pioneer for the multi-elements and multi-disciplines innovative research on the ecosystem that integrates soil, water, atmosphere, vegetation, and animal communities.

2017 年 8 月 21 日，由中国生态学学会主办的第 12 届国际生态学大会在北京国家会议中心开幕。当日下午，中心牵头组织召集了岩溶生态分会场——"岩溶生态系统：从全球到局部的意义、退化及恢复"，吸引了 60 余名国内外代表参加了会议。来自加拿大湖首大学、中国科学院华南植物园、北京林业大学等机构的 8 名学者围绕岩溶生态展开学术研讨，报告涉及岩溶区石漠化综合治理模式、岩溶生态系统服务功能评价、中国西南岩溶断陷盆地石漠化区遥感图像解译、岩溶生态与旅游、岩溶植被根系与生态水资源利用、岩溶生态系统石灰土特征、岩溶峰丛洼地水土保持机理、土壤养分对于退化岩溶区植被恢复的响应。本次岩溶生态分会场的成功召开，是生态学研究领域对于岩溶系统生态环境效应愈发重视的体现，中心将逐渐引领开展集土壤、水、大气、植被、动物群落等多要素、多学科于一体的生态系统创新研究。

The Karst Ecology Session of the 12th International Congress of Ecology
第12届国际生态学大会岩溶生态分会场现场

4. The 1st Hydrogeology and Environmental Geology Forum for Lancang-Mekong Countries

首届澜湄国家水文地质环境地质论坛

The 1st Hydrogeology and Environmental Geology Forum for Lancang-Mekong Countries was held in Beijing on 22 March 2018. As a component of the "Geosciences Cooperation Forum of Lancang-Mekong Countries", the forum aims to discuss current and future cooperation on hydrogeology and environmental geology in Lancang-Mekong countries. The forum was sponsored by Lancang-Mekong Cooperation China Secretariat and China Geological Survey, organized by IRCK/IKG, and co-organized by the Institute of Hydrogeology and Environmental Geology (IHEG) under Chinese Academy of Geological Sciences, with about 100 attendees from China and other countries attended.

2018年3月22日，首届澜湄国家水文地质环境地质论坛在北京召开，本论坛为"澜湄国家地学合作论坛"系列活动之一，旨在探讨澜湄地区水文地质环境地质合作现状与未来发展。论坛由澜湄合作中国秘书处和中国地质调查局组织，国际岩溶中心连同中国地质科学院水文地质环境地质研究所共同承办，中外嘉宾近百人参会。

During the forum, Mr. Sieng Sotham, then deputy director-general of the General Department of Mineral Resources of Cambodia; Mr. Vannaxay Xaysompheng, the geologist from the Department of Geology and Mineral Resources of the Ministry of Energy and Mines of Laos; Mr. Kyaw Kyaw Ohn, the deputy section chief of Geological Survey and Mineral Exploration Bureau under Ministry of Natural Resources and Environmental Conservation of Myanmar; Mr. Vorakit Krawchan, the geologist from the Mineral Resources Bureau under the Ministry of Natural Resources and Environment of Thailand, and Mr. Do Huy Cuong, the director of Institute of Marine Geology and Geophysics of Vietnam Academy of Science and Technology introduced hydrogeology and environmental geology research in their countries and their outlook for future cooperation. Prof. Zhang Fawang, executive deputy director of IKG, and Prof.Cheng Yanpei from IHEG made wonderful reports. Prof. Zhang expressed his willingness to consolidate the collaboration platform on hydrogeology and environmental geology for Lancang-Mekong countries through multi-lateral cooperation, scientists' exchange, training, and joint mapping on hydrogeology and environmental geology.

论坛期间，柬埔寨矿产资源总局的辛·宋哈姆副局长、老挝能源矿产部地质与矿产局的地质工程师万纳祥·先萨恒、缅甸自然资源与环境保护部地质调查与矿产勘查局的副处长玖玖吴、泰国自然资源与环境部矿产资源局的地质工程师乌拉基·克朗查、越南科学院海洋地质与地球物理研究所所长杜会强分别对本国的水文地质环境地质研究情况及今后的合作展望做了介绍。岩溶地质研究所常务副所长张发旺研究员、中国地质调查局水文地质环境地质研究所程彦培教授分别做了精彩的报告，张发旺研究员提出希望通过多双边合作，巩固澜湄水文地质环境地质跨境合作平台，开展澜湄地区水文地质环境地质人员交流与人才培养，合作开展水文地质环境地质编图。

The hydrogeology and Environmental Geology Forum for Lancang-Mekong Countries
澜湄国家水文地质环境地质论坛

5. The Academic Seminar of Task Group on Climate Change
应对气候变化团队学术研讨会

On 7 December 2019, the Academic Seminar of Task Group on Climate Change under China Geological Survey (CGS) was held in Guilin, Guangxi. At the meeting, IRCK introduced "The Preliminary Work Plan for Task Group on Climate Change under China Geological Survey", concerning 9 research fields. Each team introduced its research progress and work plan in the fields like carbon cycle in karst critical zones, risk assessment of geological disasters in townships, ecological and geological environment on Qinghai-Tibet Plateau, coastal wetland survey, geological storage of carbon dioxide, etc. All the participants proposed their suggestions about the top-level design, the general framework, and the work patterns of the task group to strengthen teamwork, and the work for national strategies to address global climate change.

2019年12月7日，中心在广西桂林举办"中国地质调查局应对气候变化团队学术研讨会"。会上，中心做了题为"中国地质调查局应对气候变化团队工作方案的初步思考"的报告，对九个方向工作提出了初步的规划和建议。团队各研究方向分别介绍了岩溶关键带碳循环、乡镇地质灾害风险评价、青藏高原生态地质环境、滨海湿地调查、二氧化碳地质储存等方面的研究成果和工作计划。与会人员围绕中国地质调查局应对气候变化调查研究工作顶层设计、工作框架、工作方式及下一步工作积极发言，就加强团队合作，服务国家应对气候变化战略问题建言献策。

The Academic Seminar of Task Group on Climate Change, CGS
中国地质调查局应对气候变化团队学术研讨会会议现场

4.1.2 Active participation in important international events for latest karst progress
积极参与各类重要国际会议，充分获取最新学科信息

4.1.2.1 The 35th International Geological Congress
第 35 届国际地质大会

IRCK delegation with Prof. Jiang Zhongcheng, Prof. Lei Mingtang, and Prof. Zhang Cheng as the delegates attended the 35th International Geological Congress held in Cape Town, South Africa from August 25 to September 3, 2016. On August 30, Prof. Jiang and Prof. Zhang presided over the session on "Karst river biogeochemical process and river basins recharge" jointly; moreover, they made poster on "Comparison of atmospheric CO_2 consumption by carbonate rocks weathering and that by silicate rocks weathering in the Pearl River Basin, China", and "Diel cycling, flux, and fate of HCO_3^- in a typical karst spring-fed stream of southwestern China". Prof. Lei orally shared a presentation titled as "Karst collapse occurrences and investigation in China" in the session of "Karst problem: identification and remediation". In addition, IRCK delegation visited the South African Geological Survey and the University of the Western Cape to exchange research results and enhance the international cooperation.

2016 年 8 月 25 日~9 月 3 日，中心代表团蒋忠诚、雷明堂、章程一行 3 人赴南非开普敦参加了第 35 届国际地质大会。会议期间，蒋忠诚研究员与章程研究员主持了"岩溶河流生物地球化学过程与流域补给"专题研讨会，并分别做了题为"中国珠江流域碳酸盐岩与硅酸盐岩风化与大气 CO_2 汇对比研究"的展板交流，以及题为"中国西南典型岩溶河流溶解无机碳昼夜循环与通量"的口头报告；雷明堂研究员则在"岩溶问题：识别与修复"专题会场做了题为"中国岩溶塌陷的分布与调查"的口头报告。其间，中心代表团访问了南非地质调查局及西开普大学，宣传中心研究成果，为国际岩溶研究中心拓展国际合作领域做出贡献。

第四章 国际交流与培训

上图 /Top

Prof. Jiang Zhongcheng introduced carbon cycle effects in the Pearl River Basin to the attendees
蒋忠诚研究员向参会代表介绍珠江流域岩溶碳循环进展

下图 /Bottom

Prof. Jiang Zhongcheng and Prof. Zhang Cheng visited the University of the Western Cape
蒋忠诚和章程研究员访问西开普大学

国际岩溶研究中心第二个六年历程

4.1.2.2 The 17th International Congress of Speleology
第 17 届国际洞穴大会

From 23 to 29 July 2017, an IRCK delegation led by Prof. Jiang Zhongcheng (GB-II member) visited Sydney of Australia to attend the 17th International Congress of Speleology (ICS). During the congress, they had a meeting with the former president, incumbent president, and secretary-general of the International Union of Speleology (UIS), as well as other international experts on karst speleology. At the meeting, Prof. Jiang introduced IRCK's intent to cooperate with UIS in the following three aspects: ① IRCK hopes to cooperate with the UIS experts on the propose of application of the Technical Committee on Karst under the International Organization for Standardization (ISO) by the Standardization Administration of the People's Republic of China (SAC); ② IRCK proposed an International Big Scientific Plan on "Global Karst" in 2016, for which IRCK is calling on closer collaboration with UIS in the fields like establishing an international monitoring network on karst critical zones, and a global karst caves database and information platform; and ③ IRCK is hoping to invite more UIS specialists to support karst-related UGGps applications. The intent was strongly agreed by all the attendees, indicated their full support to the related work.

2017 年 7 月 23～29 日，中心理事蒋忠诚研究员率领代表团赴澳大利亚悉尼参加第 17 届国际洞穴大会。其间，与国际洞穴联合会（UIS）前任、现任主席，秘书长和国际岩溶洞穴专家举行了专题座谈会。蒋忠诚理事向国际洞穴联合会的专家介绍了中心在国际合作领域的进展情况，主要包括三个方面：一是中国国家标准委员会拟向国际标准化组织（ISO）申请成立岩溶相关国际标准技术委员，邀请国际洞穴联合会专家开展合作；二是中心正在牵头实施"全球岩溶"国际大科学计划，可与国际洞穴联合会合作建设全球岩溶环境关键带监测站网、全球岩溶洞穴数据库和信息平台；三是希望在申报岩溶相关的世界地质公园期间获得国际洞穴联合会专家支持。对此与会专家均予以高度认同，并明确表示将全力支持相关活动的开展。

第四章 国际交流与培训 247

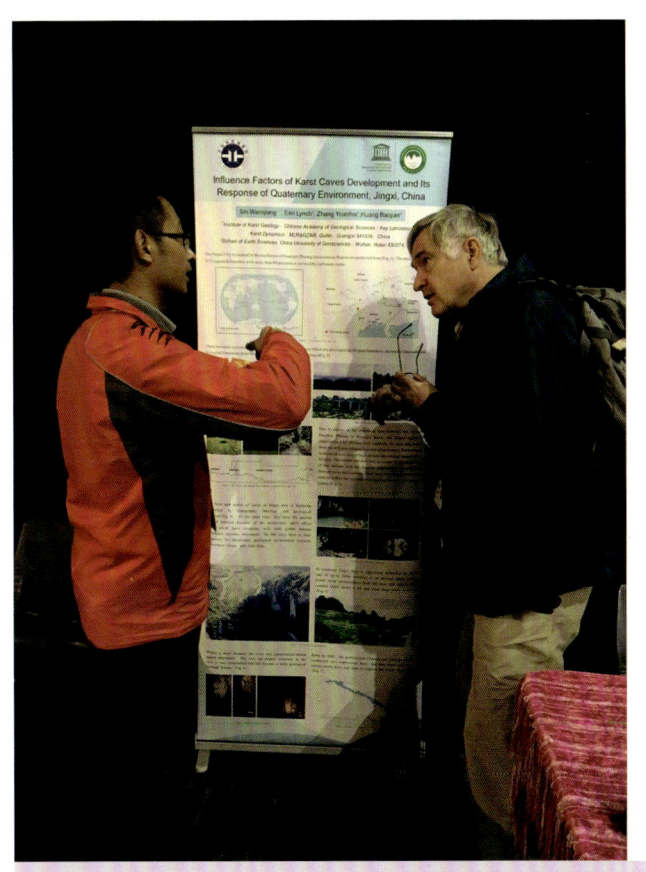

Assistant professors Shi Wenqiang (left) and Yang Lichao (right) from IRCK introduced posters at the congress
中心科技人员史文强（左）、杨利超（右）做展板介绍

4.1.2.3 The 45th International Association of Hydrogeologists Congress
第 45 届国际水文地质学家协会大会

From September 9 to 15, 2018, a 6-member delegation from IRCK visited the Republic of Korea (Daejeon) to attend the 45th IAH Congress . The congress set 9 major topics with about 45 subjects. IRCK hosted the thematic session on "Critical zone in karst system" under the seventh topic "Advances in karst and fractured-rock hydrogeology". IRCK delegation had in-depth exchanges with different experts on groundwater numerical simulation of karst fractured aquifers, aquifers protection and utilization, as well as karst critical zones, inviting international peers to implement the International Big Scientific Plan on "Global Karst" jointly through better collaboration and exchange.

2018 年 9 月 9~15 日，中心代表 6 人赴韩国大田参加 2018 年国际水文地质学家协会大会。本次大会共设置了 9 大项约 45 个专题，中心召集了第七主题"岩溶及裂隙岩体水文地质进展"中的"岩溶系统关键带"专题会议。中心代表与国内外专家就岩溶裂隙含水层地下水数值模拟、含水层保护利用及岩溶关键带问题进行了深入的交流，呼吁国际同行共同开展"全球岩溶"国际大科学计划，加强岩溶领域的国际合作与交流。

左图 /Left
Prof. Lei Mingtang: Study on mechanism of karst collapse induced by extreme climate events
雷明堂做题为"极端降雨诱发岩溶塌陷机理研究"的报告

右图 /Right
Prof. Meng Yan: Chemical characteristics of groundwater indicate the soil cave growth
蒙彦做题为"地下水化学特性指示的岩溶土洞发育规律"的报告

4.1.2.4 The 15th Multi-disciplinary Conference on Karst Collapse and Karst Engineering Environmental Effects
第 15 届岩溶塌陷与岩溶工程环境作用多学科交叉会议

On April 1–7, 2018, IRCK delegation composed of Dr. Jia Long and Dr. Pan Zongyuan visited the United States to attend the 15th Multi-disciplinary Conference on Karst Collapse and Karst Engineering Environmental Effects. The biennial conference, co-organized by the National Cave and Karst Research Institute (the US) and the Karst Waters Institute (the US), is an important platform for communication, sharing, and display. The delegates made two poster presentations: "Study on early recognition methods of cover-collapse sinkhole in China" (Jia Long) and "Research on monitoring and early warning technology for sinkholes" (Pan Zongyuan).

2018 年 4 月 1~7 日，中心贾龙博士、潘宗源博士 2 人赴美国参加第 15 届岩溶塌陷与岩溶工程环境作用多学科交叉会议。本次会议由美国国家洞穴与岩溶研究所和岩溶水研究所联合组织，该会议每两年举办一次，是全世界岩溶学及其相关领域科学家开展学术交流、经验分享和成果展示的重要平台。中心代表团分别以"中国关于岩溶塌陷早期识别的研究"（贾龙）和"岩溶塌陷监测预警研究"（潘宗源）制作了两块展板。

IRCK group at the conference
中心代表团合影

4.1.2.5 Serial meetings of Coordinating Committee for Geosciences Programs in East and Southeast Asia (CCOP)
东亚东南亚地学计划协调委员会（CCOP）系列会议

1. CCOP Groundwater Project Seminar in Vietnam
越南 CCOP 地下水培训项目研讨会

Prof. Jiang Zhongcheng, the representative of IRCK visited Vietnam to attend CCOP Groundwater Project Seminar from March 16 to 18, 2016. It was the second meeting of this project led by hydrogeologist Dr. Youhei Uchida of the Japan Geological Survey and jointly organized by CCOP, the Japan Geological Survey, and the Vietnam Water Resources Planning and Research Center. The meeting focused on the existing problems and countermeasures of groundwater in CCOP Member Countries, as well as the work plan and outcomes of this project in 2016 and 2017. IRCK, cooperating with IHEG, prepared and reported "China groundwater issues and management", focusing on the progress of groundwater investigation and research, as well as utilization and management in China, exchanged about groundwater information integration and evaluation, and clarified China's favorable impact on the international river—Red River. This meeting aroused IRCK's attention to the cooperation and publicity of transboundary rivers.

2016 年 3 月 16~18 日，中心理事蒋忠诚研究员赴越南河内参加了 CCOP 地下水培训项目会议。本次会议为日本地调局水文地质专家 Youhei Uchida 博士负责的地下水项目的第二次会议，由 CCOP、日本地调局和越南水资源规划和调查中心联合召开。会议围绕 CCOP 成员国地下水目前存在的问题和对策，以及项目今明两年的工作计划和成果资料展开讨论。中心联合中国地质科学院水文地质环境地质研究所负责编制和汇报"中国地下水问题与行动对策"，展示了中国地下水的调查研究和开发治理的进展，学习交流了地下水信息集成与评价的先进经验，辨明了中国对红河国际河流的有利影响。通过参加本次会议，中心加强了跨境河流方面的合作研究与宣传。

The delegation visited the Red River for its hydrological conditions and water resources
与会代表考察红河水文水资源情况

2. "Climate Change and Groundwater Resources in Mekong River Basin" Seminar
"湄公河流域气候变化与地下水资源"研讨会

On June 1-2, 2016, Prof. Zhang Cheng, secretary-general of IRCK, attended the "Climate Change and Groundwater Resources in Mekong River Basin" Seminar held in Cambodia. With CCOP, Korea Institute of Geoscience and Mineral Resources (KIGAM), UNESCO Bangkok Office, and Cambodia Mineral Resources Office of General Department of Mineral Resources (GDMR) as the organizers, the seminar aimed to enhance the understanding of the impacts from climate change to groundwater, strengthen exchanges and cooperation among countries in the Mekong River Basin region, and share experience in mitigating climate change impacts and groundwater management for finding out corresponding strategies and measures. At the invitation of the CCOP Secretariat, IRCK, cooperated with IHEG, made a report as "Climate change and groundwater resources in China", covering the hydrogeological characteristics of north and south China, the impact on groundwater by climate change, as well as the strategies and actions to cope with drought. A total of 31 participants from CCOP Member Countries, UNESCO Bangkok Office, and USGS attended.

2016年6月1~2日，中心秘书长章程研究员赴柬埔寨参加"湄公河流域气候变化与地下水资源"研讨会。本次研讨会由CCOP、韩国地球科学与矿产资源研究所（KIGAM）、UNESCO曼谷办事处、柬埔寨矿产能源部矿产资源总局（GDMR）联合主办。研讨会旨在提升气候变化对地下水资源影响的认识，加强湄公河流域国家间的交流与合作，分享缓解气候变化影响实践与地下水管理经验，共同探讨应对策略与途径。受CCOP秘书处的邀请，中心联合中国地质科学院水文地质环境地质研究所共同做了题为"中国地下水资源特点及气候变化的影响"的报告，介绍了中国南北方水文地质特点、气候变化对地下水的影响、应对干旱策略与行动及取得的成效与经验。CCOP成员国及UNESCO曼谷办事处、美国地调局代表共31人参加了会议。

A representative made a country report during the seminar
参会代表做国家报告

3. The 53rd CCOP Annual Session and Academic Symposium
第 53 届 CCOP 年会暨学术研讨会

On 15-20 October 2017, Prof. Zhang Cheng, secretary-general of IRCK, visited the Philippines to attend the 53rd CCOP Annual Session and Academic Symposium. The session, setting a theme of "The Role of Geosciences in Safeguarding Environment", was sponsored by CCOP and organized by the Mines and Geosciences Bureau of the Department of Environment and Natural Resources of the Philippines, and the Energy Resource Development Bureau of the Department of Energy of the Philippines, and CCOP Technical Secretariat jointly. During the session, Prof. Zhang delivered an oral presentation on "Diurnal changes of DIC and EC of rivers in subtropical karst areas", and introduced the progress of IGCP 661 led by IRCK.

2017 年 10 月 15~20 日，中心秘书长章程研究员赴菲律宾参加第 53 届 CCOP 年会暨学术研讨会。本次会议主题为"地球科学如何为环境保驾护航"，由 CCOP 主办，菲律宾自然资源与环境部矿产与地球科学局、菲律宾能源部能源资源发展局、CCOP 技秘书处联合承办。章程研究员做了题为"亚热带岩溶区河流溶解无机碳与电导率的昼夜变化"的口头报告，并宣传了中心主持的 IGCP 661 项目的进展情况。

Prof. Zhang Cheng made a presentation
章程研究员做报告

4. The 56th CCOP Annual Session
第 56 届 CCOP 年会

On 3-4 November 2020, the 56th CCOP Annual Session was held virtually. The representatives from 12 CCOP member states, 9 cooperating countries (e.g. Australia, Belgium, Canada, Denmark, etc.), as well as 9 international organizations (e.g. UNESCO, the International Union of Geological Sciences (IUGS), the Young Earth Scientists (YES), etc.). Two representatives from IRCK attended the meeting and learned about the research progress and future trends of different countries. *The China Annual Report on Karst Geology* compiled by IRCK is an important part of the Country Report China, especially the research progress of karst carbon cycle and global climate change, water resources development and management, landscape resources survey and conservation, as well as rocky desertification control, which made great contributions to the UN 2030 Agenda, also had a great significance to regional cooperation organized by CCOP.

2020年11月3~4日，东亚东南亚地学计划协调委员会（CCOP）第56届年会在线上召开，12个成员国和澳大利亚、比利时、加拿大、丹麦等9个合作国，以及联合国教科文组织、国际地质科学联合会、国际青年地质学家协会等9个国际合作组织代表参会。中心两名代表在线参会，学习了解了各国在地球科学领域的研究成果及工作动态。中心编制的岩溶地质工作年报作为中国2020年度CCOP国家报告的重要组成部分。其中，岩溶碳循环与全球气候变化研究、水资源开发管理、景观资源调查与保育、石漠化治理等工作成果不仅有效地服务了联合国2030年可持续发展议程，并且对于CCOP组织开展区域合作具有重要意义。

China Country Report 2020
中国 2020 年度 CCOP 国家报告

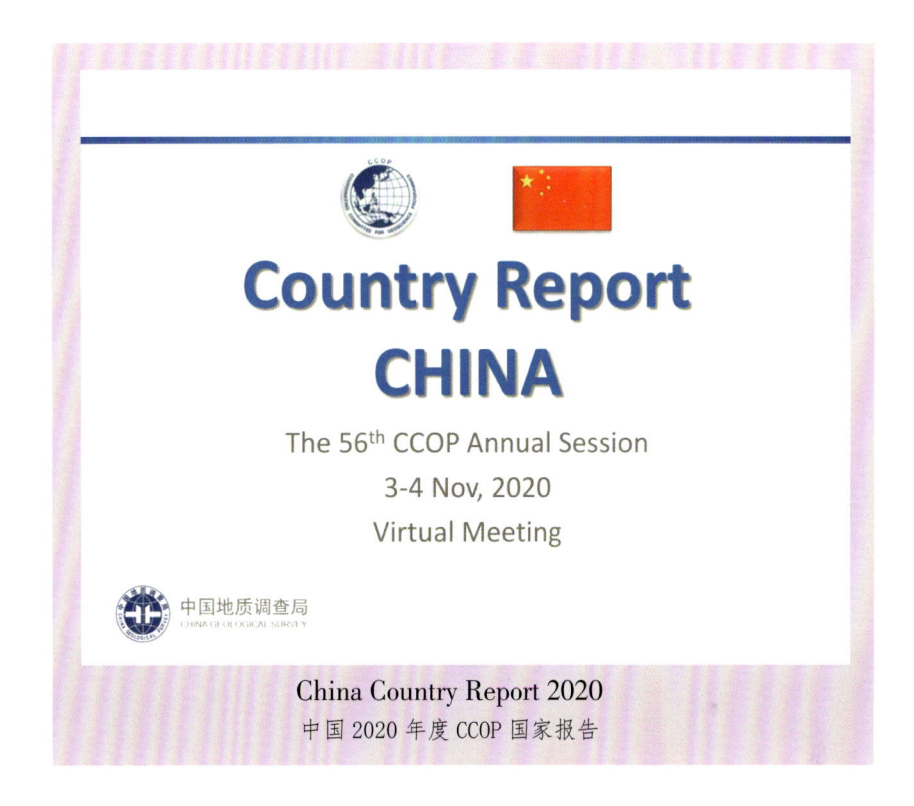

4.1.2.6 Other international academic conferences
其他各类国际学术会议

From 2016 to 2021, IRCK representatives took part in 19 virtual or on-site international academic conferences. Through multi-channel exchange and experience sharing, IRCK harvested the latest progress of karst research, providing a firm foundation for possible cooperative projects in future.

2016~2021 年，中心共派人参加了 19 次线上线下国际学术会议，通过多渠道交流分享经验，为中心掌握世界岩溶研究最新进展，开展未来项目层面的合作奠定了坚实基础。

List of international academic conferences attended by IRCK staff
中心科技人员参加其他国际学术会议一览表

No. 序号	Conference 会议名称	Time 会议时间	Venue 会议地点
1	The 5th European Speleological Congress 欧洲第 5 届洞穴会议	Aug. 13–20, 2016 2016 年 8 月 13~20 日	Yorkshire, the UK 英国约克郡
2	US Geological Survey Karst Interest Group Workshop 美国地质调查局岩溶工作组会议	May 16–18, 2017 2017 年 5 月 16~18 日	San Antonio, the USA 美国圣安东尼奥
3	Karst Record 8 (KR-8) Conference 第 8 次岩溶记录国际会议	May 21–24, 2017 2017 年 5 月 21~24 日	Austin, the USA 美国奥斯汀
4	The 11th Asian Regional Conference of IAEG 第 11 届亚洲区域工程地质大会	Nov. 28–30, 2017 2017 年 11 月 28~30 日	Kathmandu, Nepal 尼泊尔加德满都
5	The international training course and field seminar "Characterization and Engineering of Karst" and Symposium "Karst 2018—Expect the Unexpected" "岩溶含水层特征与工程"国际培训班及"岩溶 2018：期待未知"国际培训班	May 27–June 9, 2018 2018 年 5 月 27 日 ~6 月 9 日	Trebinje, Bosnia and Herzegovina 波黑特雷比涅
6	The 15th Annual Meeting of Asia-Oceania Geosciences Society (AOGS) 第 15 届亚洲大洋洲地球科学学会（AOGS）年会	June 3–8, 2018 2018 年 6 月 3~8 日	Honolulu, the USA 美国火奴鲁鲁
7	Resources for Future Generation (RFG) 2018 Conference 2018 后世资源会议	June 16–21, 2018 2018 年 6 月 16~21 日	Vancouver, Canada 加拿大温哥华

List of international academic conferences attended by IRCK staff
中心科技人员参加其他国际学术会议一览表

（续表）

No. 序号	Conference 会议名称	Time 会议时间	Venue 会议地点
8	EuroKarst 欧洲岩溶会议	July 2–6, 2018 2018年7月2~6日	Basancon, France 法国贝桑松
9	The 4th International Conference on Water Resources and Environment 第4届水资源与环境国际会议	July 17–21, 2018 2018年7月17~21日	Taiwan, China 中国台湾
10	2018 Geological Society of America (GSA) Annual Meeting 2018美国地质学会年会	Nov. 2–9, 2018 2018年11月2~9日	Boris, the USA 美国波利斯
11	European Geosciences Union General Assembly 2019 2019欧洲地球科学联合会年会	April 7–12, 2019 2019年4月7~12日	Vienna, Austria 奥地利维也纳
12	The 10th International Symposium on Managed Aquifer Recharge 第10届含水层和补给管理国际研讨会	May 19–25, 2019 2019年5月19~25日	Madrid, Spain 西班牙马德里
13	The 16th International Symposium on Water-Rock Interaction (WRI-16) and the 13th International Symposium on Applied Isotope Geochemistry (AIG -13) 第16届水岩相互作用国际研讨会暨第13届应用同位素地球化学国际研讨会	July 21–26, 2019 2019年7月21~26日	Tomsk, Russia 俄罗斯托木斯克
14	Goldschmidt Conference 2019 2019戈尔德施密特国际地球化学年会	Aug. 18–23, 2019 2019年8月18~23日	Barcelona, Spain 西班牙巴塞罗那

List of international academic conferences attended by IRCK staff
中心科技人员参加其他国际学术会议一览表

（续表）

No. 序号	Conference 会议名称	Time 会议时间	Venue 会议地点
15	Science to Enable and Empower Asia Pacific for Sustainable Development Goals (SEEAP for SDGs) II 第 2 届科学促进亚太地区实现可持续发展目标会议	Sep. 17–19, 2019 2019 年 9 月 17~19 日	Jakarta, Indonesia 印度尼西亚雅加达
16	Protection of Fragile Karst Resources 脆弱岩溶资源保护大会	Aug. 18–20, 2020 2020 年 8 月 18~20 日	Bowling Green, the USA (virtually) 美国鲍灵格林（线上）
17	Experts Dialogue on SETI Priorities and Implementation Means Asia-Pacific on line Regional Consultation 联合国教科文组织亚太区域专家咨询及科学工程技术创新机制会议	Sep. 1, 2020 2020 年 9 月 1 日	Jakarta, Indonesia (virtually) 印度尼西亚雅加达（线上）
18	A Week with Karst "岩溶一周"在线活动	June 1–7, 2021 2021 年 6 月 1~7 日	Belgrade, Serbia (virtually) 塞尔维亚贝尔格莱德（线上）
19	The 16th Regional Geoscience Congress of Southeast Asian (GEOSEA XVI) 第 16 届东南亚区域地球科学大会	Dec. 6–8, 2021 2021 年 12 月 6~8 日	Manila, Philippines (virtually) 菲律宾马尼拉（线上）

4.2 International Training Course
国际培训

4.2.1 Organize annual international training courses to provide an incubation platform for scientific and technological progress of karst-related countries around the world actively
积极组织年度国际培训，为全球岩溶相关国科技进步提供孵化平台

From 2016 to 2021, IRCK has held 8 training courses with 228 international trainees involved, benefiting 44 countries, including 154 from Asia, 30 from Africa, 24 from Europe, and 20 from America. Totally, there are 220 participants from developing countries; while the proportions of male and female are 60% and 40% respectively.

2016~2021 年中心共举办培训班 8 次，累计培训国际学员 228 人次，惠及国家 44 个，其中亚洲国家 154 人、非洲国家 30 人、欧洲国家 24 人、美洲国家 20 人；来自发展中国家的学员 220 人；男女学员占比分别为 60% 和 40%。

1. The 8th International Training Course on "Integrated Control on Rocky Desertification, Carbon Cycle and Sink in Karst System"
第八届国际培训班——岩溶石漠化综合治理、碳循环及碳汇效应

The 8th International Training Course on "Integrated Control on Rocky Desertification, Carbon Cycle and Sink in Karst System" was held in Guilin from November 13 to 27 of 2016. This training course focused on the rocky desertification treatment, carbon cycle and carbon sink effects of karst system, addressing global climate change, karst collapse monitoring and early warning, karst water resources development and utilization, karst oil and gas resources research, as well as karst landscape and cave research. It aimed to take full advantage of IRCK platform to enhance the technical and theoretical capacity for the related countries. The 15-day training course included the announcement ceremony of "Global Karst", the 1st meeting of the 2nd Academic Committee (academic exchange), lectures, field practice, and trainee assessment. It enrolled 31 trainees from 13 countries, including 18 students from 12 foreign countries like Thailand, Brazil, Slovenia, and Myanmar; meanwhile, it invited 20 experts from Indonesia, Serbia, Poland, Austria, Brazil, the United States, and China as lecturers, some of them are GB and AC members. All the trainees gave high comments to IRCK for its effectiveness as an international platform for promoting learning and exchanging. Finally, eight trainees from Brazil, Iran, Slovenia, Serbia, Macedonia, Thailand, and Zimbabwe were selected as excellent trainees after a tense but interesting trainee assessment.

2016年11月13~27日，中心第八届国际培训班——"岩溶石漠化综合治理、碳循环及碳汇效应"在桂林举办。本次培训班围绕岩溶系统中的石漠化治理、碳循环及碳汇效应、应对全球气候变化、岩溶塌陷监测预警、岩溶水资源开发利用、岩溶油气资源研究、岩溶景观与洞穴研究等内容进行课程设置，旨在提升相关国家的技术和理论水平，发挥中心平台作用。本次培训班为期15天，内容包括："全球岩溶"国际大科学计划发布仪式、第二届学术委员会第一次会议（学术交流）、室内授课、野外实习及结业评估。共招收了来自13个国家的31名学员，其中外籍学员18人，分别来自泰国、巴西、斯洛文尼亚、缅甸等12个国家；此外，邀请了来自印度尼西亚、塞尔维亚、波兰、奥地利、巴西、美国及中国的20名专家（部分为中心理事或学术委员）为培训班授课。学员们充分肯定了中心带给他们的学习和交流机会，最后通过紧张有趣的学员评估环节，来自巴西、伊朗、斯洛文尼亚、塞尔维亚、马其顿、泰国、津巴布韦的8名学员被评为了优秀学员。

Group photo of the 2016 IRCK training course
2016年中心培训班合影

Chapter 4　International Exchange and Training

左上图 /Top left
Eko Haryono (Indonesia): Carbon cycle and flux in karst
艾可·哈约诺（印度尼西亚）：岩溶区碳循环与碳通量

左下图 /Bottom left
Sasa Milanovic (Serbia): Research methods of karst systems from recharge to discharge zones (examples from Dinaric and Carpathian-Balkan karst);
萨沙·米拉诺维奇（塞尔维亚）：从补给区到排泄区岩溶系统研究方法（以第纳尔和喀尔巴阡-巴尔干岩溶为例）

右上图 /Top right
Andrzej Tyc (Poland): Application of airborne (LiDAR) and terrestrial scanning to karst studies
安德烈·泰克（波兰）：航空扫描（LiDAR）与陆地扫描系统在岩溶研究领域的应用

右下图 /Bottom right
Petar Milanovic (Serbia): Aeration zone in karst properties and investigations
皮特·米拉诺维奇（塞尔维亚）：岩溶区包气带特征及调查

第四章 国际交流与培训

从左至右、从上至下依次为

From left to right, top to bottom are

Yuan Daoxian (China): Hydrogeology and environmental geology issues in karst of China

袁道先（中国）：中国岩溶的水文地质和环境地质问题

Jut Wynne (the USA): Human-induced change and the effects on cave ecosystems—A case study of China

约特·怀尼（美国）：人类行为对洞穴生态系统的影响及效应——以中国研究为例

George Veni (the USA): The impacts and management of climate change on karst hydrology and ecosystems

乔治·维纳（美国）：气候变化对岩溶水文和生态系统的影响和管理

Augusto Auler (Brazil): Paleoenvironmental changes in tropical areas derided from cave deposits: Examples from Brazil

奥古斯托·奥乐（巴西）：利用热带洞穴沉积物重建古环境变化——巴西案例分析

Li Wei (China): Carbon concentration mechanisms in aquatic macrophytes

李为（中国）：大型水生植物碳聚集机制

Chapter 4　International Exchange and Training

上图 /Top
Prof. Guo Fang introduced the Yaji Experimental Site
郭芳研究员介绍丫吉试验场功能

中图 /Middle
Prof. Yi Lianxing introduced the Haiyang–Zhaidi underground river system
易连兴研究员介绍海洋－寨底地下河系统

下图 /Bottom
Group photo at Maocun Experimental Site
毛村野外基地合影

Granting certificates to the trainees
颁发结业证书

2. The 9th International Training Course on "Karst Hydrogeology, Karst Environmental Geology and Karst Ecological Geology" & Training Course on Groundwater and Environmental Geology Investigation and Evaluation for ASEAN Member Countries

第九届国际培训班——岩溶水文地质、环境地质、生态地质暨东盟地下水资源与环境地质调查评价技术培训

On 17-31 July 2017, the 9th International Training Course on "Karst Hydrogeology, Karst Environmental Geology and Karst Ecological Geology" & Training Course on Groundwater and Environmental Geology Investigation and Evaluation for ASEAN member states convened in Kunming, Yunnan. With the congratulatory letter sent by Mr. Hans Thulstrup, senior programme specialist at UNESCO Regional Science Bureau for Asia and the Pacific, the training course was organized successfully.

2017年7月17~31日，"岩溶水文地质、环境地质、生态地质"国际培训班暨"东盟地下水资源与环境地质调查评价技术培训"在云南昆明召开。联合国教科文组织亚太科学局高级项目专员汉斯·图尔斯特鲁普先生专门发来贺信。

The training course was jointly sponsored by IRCK's operational fund and Asian regional cooperation fund. Nearly 20 lecturers from the United States, Serbia, Slovenia, and China were invited, with more than 60 trainees enrolled from Brazil, China, Egypt, Hungary, Iran, Mongolia, etc. With the most participants and participating countries, its scale had ever been the largest one since the interception of IRCK, it's also the first time for IRCK to integrate regional cooperation with global promotion, featured by targeting training to different research status of different countries (regions). The trainees/groups from Hungary, Iran, Brazil, Slovenia, Myanmar, Thailand (the participants from the Department of Groundwater Resources), and China (the participants from Southwest University) were evaluated as excellent trainees/groups.

本届培训班结合了中心承担的亚洲区域合作专项资金项目共同举办，共邀请了来自美国、塞尔维亚、斯洛文尼亚和中国的近20位教员，招收了来自巴西、中国、埃及、匈牙利、伊朗、蒙古等国家的60余位学员。本届培训班是国际岩溶研究中心成立以来规模

最大、参与国家和人数最多的培训班，也是中心首次尝试将区域合作与全球传播予以有机结合，并展开针对不同区域研究现状的国际培训。来自匈牙利、伊朗、巴西、斯洛文尼亚、缅甸、泰国地下水资源厅，以及中国西南大学的学员或团队被评为优秀学员/团队。

The training course was highly recognized by all the participants, and nearly all the trainees expressed their intent to take part in more training events organized by IRCK in future.

本次培训班获得了各方高度认可，学员们都表示希望能有更多机会参加由中心举办的国际培训。

Hu Xiaonong (the USA): Monitoring, experiments and modeling of karst hydro-dynamics
胡晓农（美国）：岩溶水动力学——观测、试验和建模

Jonathan Arthur (the USA): A review of the hydrogeology and ecosystem restoration plans for the Edwards Aquifer, Texas
乔纳森·亚瑟（美国）：得克萨斯爱德华含水层水文地质和生态系统恢复规划

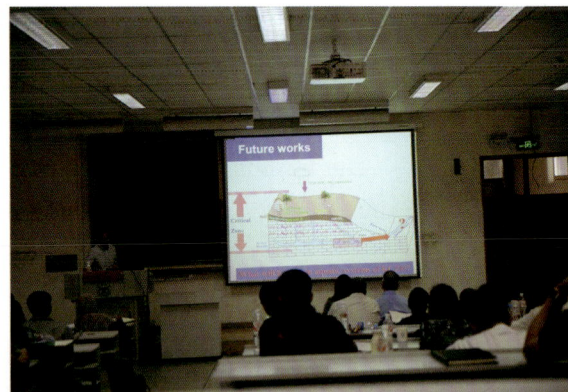

左上图 /Top left
Sasa Milanovic (Serbia): Hydrogeological engineering of karst aquifer—Problems and solutions
萨沙·米拉诺维奇（塞尔维亚）：岩溶含水层水文地质工程——问题与解决路径

左下图 /Bottom left
Tian Tingshan (China): Emergency prevention of geological disaster
田廷山（中国）：地质灾害应急预防

右上图 /Top /right
Jiang Yongjun (China): Carbon-nitrogen coupling cycle in karst critical zone
蒋勇军（中国）：岩溶关键带C-N耦合循环

右下图 /Bottom right
Tang Jiansheng (China): Hydrogeological survey in karst area (1:50,000)
唐建生（中国）：岩溶区水文地质调查（1:50000）

上图 /Top

Wang Hongmei (China): Microbial communities in karst area and their functions

王红梅（中国）：岩溶区微生物群落及其功能

中图 /Middle

Yi Zhijun (Permanent mission of China to UNESCO): UNESCO's Intergovernmental Programme: Concept Implementation

易志军（中国驻教科文组织常驻团）：教科文组织政府间项目：概念执行

下图 /Bottom

Yuan Daoxian (China): Discussions on the critical zones in the karst areas of China

袁道先（中国）：中国岩溶关键带问题探讨

左上图 /Top left
Field practice of hydrogeological and environmental geological investigation in Dabanqiao Town
大板桥镇水文地质和环境地质调查野外实习

左下图 /Bottom left
Field practice of hydrogeological and environmental geological investigation of rocky desertification control in Luxi Basin
泸西盆地石漠化治理水文地质环境地质调查野外实习

右上图 /Top right
Visit Yunnan Shilin Karst Geology Museum
参观云南石林岩溶地质博物馆

右下图 /Bottom right
Field investigation of water storage mechanism of Wulichong Reservoir
五里冲水库蓄水机理野外实习

第四章 国际交流与培训

Closing ceremony of the training course
培训班结业典礼

3. The 10th International Training Course on "Karst Landscape and Karst Ecological Geology"

第十届国际培训班——岩溶景观及岩溶生态地质

On 13-26 November 2018, the 10th International Training Course on "Karst Landscape and Karst Ecological Geology" was held in Nanning, Xiangxi, and Guilin. It combined the 9th China-ASEAN Mining Cooperation Forum (Mining Cooperation Forum), the 2nd International Symposium on Karst of Xiangxi Geopark (Xiangxi Symposium), and China-ASEAN International Forum on Sustainable Development and Innovative Cooperation & China (Guilin) High-level International Forum on Health Tourism (Sustainable Development Forum), creating a diversified, plentiful, and fruitful training. A total of 10 lecturers were invited from 9 countries like the US, Austria, Brazil, Indonesia, etc.; and 25 trainees were enrolled from 16 countries like Egypt, Ethiopia, Brazil, etc.

2018年11月13~26日，中心在南宁、湘西、桂林三地召开了第十届国际培训班——"岩溶景观及岩溶生态地质"。本次培训班结合了第九届中国-东盟矿业合作论坛、湘西地质公园第二届喀斯特国际学术研讨会及中国-东盟可持续发展创新合作论坛暨中国（桂林）国际健康旅游高端论坛，形式多样，内容丰富，成果丰硕。参加此次培训班的教学员共计35人，其中教员10人，分别来自美国、巴西、印度尼西亚等9个国家；学员25人，分别来自埃及、埃塞俄比亚、巴西等16个国家。

The trainees learned more about the regional cooperation progress on mining through the Mining Cooperation Forum, got a better understanding of the special karst landscape evolution (e.g. Red Stone Forest) in Xiangxi Geopark through the Xiangxi Symposium, and knew more about how to utilize karst resources to support the sustainable development of Guilin through the Sustainable Development Forum. After the trainee assessment, a total of 5 excellent trainees were recognized who are from Slovenia, Poland, Egypt, Indonesia, and Brazil.

培训班学员在广西南宁参加中国-东盟矿业合作论坛，了解区域矿业合作研究进展；赴湖南湘西参加湘西地质公园第二届喀斯特国际学术研讨会，掌握湘西地质公园岩溶特色，了解湘西地区红石林等特色景观的发育特征；在广西桂林参加中国–东盟可持续发展创新合作论坛暨中国(桂林)国际健康旅游高端论坛，学习如何利用岩溶特色资源支撑桂林市可持续发展。最后，本届培训班评选出了5位优秀学员，分别来自斯洛文尼亚、波兰、埃及、印度尼西亚和巴西。

左图/Left
Andrej Tyc (Poland): Collapse dolines caused by hypogene karstification
安德烈·泰克（波兰）：深部岩溶作用与塌陷成因漏斗

右图/Right
Ralf Benischke (Austria): Karst groundwater monitoring targets, strategies and evaluation
拉尔夫·比尼斯尔克（奥地利）：岩溶地下河监测目标、策略与评价

Chapter 4 International Exchange and Training

左上图 /Top left

Augusto Auler (Brazil): Paleoenvironmental change in tropical areas derided from cave deposits—Examples from Brazil

奥古斯托·奥乐（巴西）：利用热带洞穴沉积物重建古气候变化——巴西案例分析

左下图 /Bottom left

Jut Wynne (the USA): Towards an iterative vulnerability assessment of cave biota: South China karst, a case study

约特·怀尼（美国）：洞穴生物群落迭代脆弱性评估：中国南方岩溶案例分析

右上图 /Top right

Zoran Stevanovic (Serbia): Engineering solutions in optimizing groundwater supply and environmental impacts indicators

佐伦·史蒂夫诺维奇（塞尔维亚）：优化地下水供给的工程方案和环境影响指标

右下图 /Bottom right

George Veni (the USA): The impacts and management of climate change on karst hydrology and ecosystems

乔治·维纳（美国）：气候变化对岩溶水文和生态系统带来的影响及其对策

第四章　国际交流与培训

上图 /Top
Yuan Daoxian (China): Karst and its environmental issues along the Silk Road
袁道先（中国）：丝绸之路沿线的岩溶环境问题

下图 /Bottom
Closing ceremony of the training course
培训班结业典礼

国际岩溶研究中心第二个六年历程

4. The 11th International Training Course on the " 'The Belt and Road' and Karst Landscape"

第十一届国际培训班——"一带一路"与岩溶景观

On Sep.16-28, 2019, the International Training Course on the "'The Belt and Road' and Karst Landscape" was successfully held in Guilin, China. IRCK invited 14 lecturers from 8 countries including Austria, Croatia, the USA, etc., and enrolled 41 trainees from 17 countries like Brazil, Poland, Russia, etc. The lectures helped the trainees learn more about the formation and evolution of karst caves, the characteristics of karst development in different areas and the reasonable exploitation and utilization of water resources in karst areas; the field study in the World Natural Heritage Site in Guilin and Yaji Karst Experimental Site enabled the trainees to know more about the formation and evolution of Guilin karst; additionally, the team assessment (the trainees were divided into 3 teams according to their expertise, e.g. Karst Hydrogeology Team, Karst Environmental Geology Team, and Karst Landform Team) not only strengthened the interactive communication between the trainees, but also stimulated their teamwork intent. Finally, the Karst Environmental Geology Team composed of trainees from Brazil, the Philippines, Jordan, Myanmar, and Thailand was evaluated as the "Excellent Team", while nine excellent trainees from Brazil, Thailand, France, the Philippines, and Iran were recognized after trainee assessment.

2019年9月16~28日，中心第十一届国际培训班——"'一带一路'与岩溶景观"国际培训班在广西桂林成功举办。本次培训班共邀请了来自奥地利、克罗地亚、美国等8个国家的14位教员，招收了来自巴西、波兰、俄罗斯等17个国家的41位学员。通过室内授课，学员们学习了岩溶洞穴景观的形成和演化规律、不同地区岩溶发育的特征及岩溶区水资源合理开发利用等内容；野外实地考察了桂林世界自然遗产地、丫吉野外试验场，充分了解了桂林不同岩溶地貌的形成和演化规律。本次培训班根据学员专业背景，将学员分为岩溶水文地质组、岩溶地貌组及岩溶环境地质组，首次将团队评估和个人评估相结合，不仅促进了学员之间的互动，也激发了学员之间的团队合作精神。经评选，由巴西、菲律宾、约旦、缅甸、泰国学员组成的岩溶环境地质组被评为优秀团队，同时，来自巴西、泰国、法国、菲律宾、伊朗的9名学员代表被评为优秀学员。

第四章　国际交流与培训

上图 /Top
Group photo of the opening ceremony of the training course
培训班开幕式合影

下图 /Bottom
Sasa Milanovic (Serbia): Speleology, cave diving investigations in karst engineering
萨沙·米拉诺维奇（塞尔维亚）：岩溶工程领域洞穴学与洞穴潜水调查

国际岩溶研究中心第二个六年历程

左上图 /Top left
Chaiporn Siripornpibu (Thailand): Status of CK management in Thailand after cave rescue
柴鹏·斯里蓬皮布尔（泰国）：洞穴营救后泰国洞穴和岩溶管理现状

左下图 /Bottom left
Harrison Pienaar (South Africa): Groundwater protection measures in South Africa
哈里森·皮纳尔（南非）：南非地下水保护措施

右上图 /Top right
Ralf Benischke (Austria): Characterastics of Alpine karst of Austria
拉尔夫·比尼斯尔克（奥地利）：奥地利高寒山区岩溶特征

右下图 /Bottom right
Chris Groves (the USA): The Mammoth Cave National Park World Heritage Site: Protecting the world's longest known cave
克里斯·葛立夫（美国）：世界自然遗产猛犸洞国家公园——保护世界上已知的最长洞穴

上图 /Top
Ognjen Bonacci (Croatia): Karst water resources—Ecohydrological aspects
欧根·伯纳茨（克罗地亚）：岩溶水资源　生态水文学研究

中图 /Middle
Yuan Daoxian (China): Karst and related hydrological & environmental issues along the Silk Road
袁道先（中国）：丝绸之路沿线岩溶水文和环境问题

下图 /Bottom
Eko Haryono (Indonesia): Methods in assessing karst morphology and karst aquifer development
艾可·哈约诺（印度尼西亚）：岩溶形态评估方法及岩溶含水层开发

Chapter 4　*International Exchange and Training*

Team discussion
学员分组讨论

Team assessment
小组答辩

第四章 国际交流与培训

左上图 /Top left
Prof. Jiang Guanghui introduced the Yaji Experimental Site
姜光辉研究员介绍丫吉试验场

右上图 /Top right
Prof. Yi Lianxing introduced the Haiyang-Zhaidi Underground River
易连兴研究员介绍海洋-寨底地下河系统

下图 /Bottom
Group photo of the closing ceremony (2019)
2019 年培训班结业合影

国际岩溶研究中心第二个六年历程

5. The 12th International Training Course on "Karst Resources, Environmental Effects and Ecological Industry"

第十二届国际培训班——岩溶资源环境与生态产业

On October 26–30 and November 9–13 2020, IRCK organized the 12th International Training Course on "Karst Resources, Environmental Effects and Ecological Industry" virtually in two phases. Due to the impact of COVID-19, this is the first time for IRCK to held a virtual training course. A total of 17 well-known karst scientists from the US, Serbia, Poland, the Philippines, and other countries were invited as lecturers, while 36 foreign trainees from 16 countries attended this training course. The training course was strongly supported by the UNESCO Beijing Office, which helped to enroll the trainees.

2020年10月26~30日、11月9~13日，岩溶中心分两期举办了第十二届国际培训班——"岩溶资源环境与生态产业"。受新冠疫情影响，这是中心首次开展线上培训，共邀请到来自美国、塞尔维亚、波兰、菲律宾等的17名著名岩溶学者担任教员进行授课，招收了来自巴西、葡萄牙、斯洛文尼亚、牙买加、菲律宾、缅甸等16个国家的36名外籍学员在线参与。本次培训班获得了联合国教科文组织驻华代表处的大力支持，协助中心完成招生。

On October 30 and November 13, the trainees made their presentations for the final assessment, with 8 trainees from Slovenia, the Philippines, Morocco, China, Brazil, Myanmar, and Iran as the excellent ones. The successful virtual training course provided a good experience for IRCK to explore more diversified training modes. The excellent reports from the lecturers and trainees enriched the data for karst research, laying a firm foundation for multilateral cooperation in future.

10月30日及11月13日，参训学员分两期进行了结业汇报，经评委评议，共选出来自斯洛文尼亚、菲律宾、摩洛哥、中国、巴西、缅甸、伊朗的8名优秀学员。此次线上培训班的顺利举办为中心探索更加多元的培训模式提供了良好经验，教员及学员的优秀报告极大地丰富了世界各国的岩溶研究素材，为未来开展多边合作奠定了坚实基础。

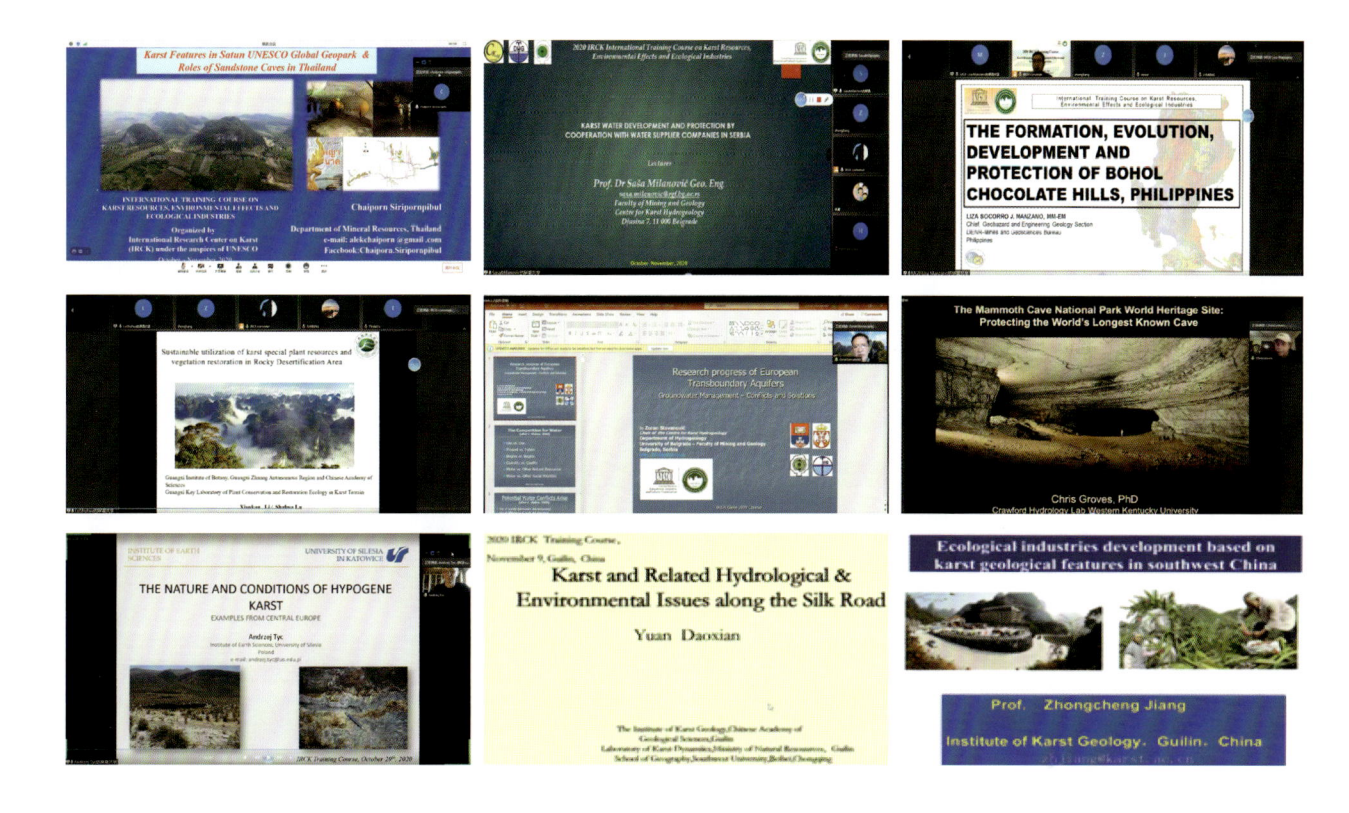

从左至右、从上至下依次为

From left to right, top to bottom are

Chaiporn Siripornpibul (Thailand): Karst features in Satun UNESCO Global Geopark & roles of sandstone caves in Thailand; Sasa Milanovic (Serbia): Karst water development and protection by cooperation with water supplier companies in Serbia; Liza Socorro J. Manzano (the Philippines): The formation, evolution, development and protection of Bohol Chocolate Hills, Philippines; Li Xiankun & Lu Shuhua (China): Sustainable utilization of karst special plant resources and vegetation restoration in rocky desertification area; Zoran Stevanovic (Serbia): Research progress of European transboundary aquifers; Chris Groves (the USA): Mammoth Cave National Park World Heritage Site: Protecting the world's longest known cave; Andrej Tyc (Poland): The nature and conditions of hypogene karst; Yuan Daoxian (China): Karst and related hydrological & environmental issues along the Silk Road; Jiang Zhongcheng (China): Ecological industries development based on karst geological features in southwest China

柴鹏·斯里蓬皮布尔（泰国）：泰国沙敦世界地质公园岩溶特征及砂岩洞穴特征；萨沙·米拉诺维奇（塞尔维亚）：塞尔维亚岩溶水资源开发与保护——与供水公司合作；丽萨·曼赞诺（菲律宾）：菲律宾薄荷岛巧克力山的成因、演化、开发与保护；李先琨和陆树华（中国）：岩溶石山区特有植物资源可持续利用和植被修复；佐伦·史蒂夫诺维奇（塞尔维亚）：欧洲跨边界含水层研究进展；克里斯·葛立夫（美国）：世界自然遗产猛犸洞国家公园——保护世界上已知的最长洞穴；安德烈·泰克（波兰）：深部岩溶的特点与条件；袁道先（中国）：丝绸之路沿线岩溶水文地质和环境地质问题；蒋忠诚（中国）：中国西南地区基于岩溶地质特征的生态产业发展

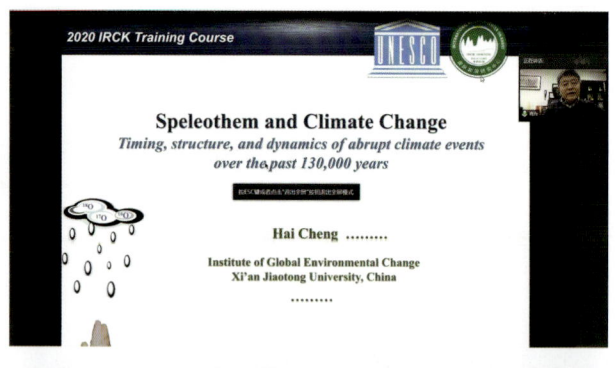

左上图 /Top left

Zheng Yuanyuan (China): UNESCO Global Geoparks in China: Activities & events

郑园园（中国）：中国的世界地质公园：大事记

左下图 /Bottom left

Cao Jianhua (China): Integrated control on karst rocky desertification in graben basin, eastern Yunnan Plateau

曹建华（中国）：云南高原东部断陷盆地岩溶石漠化综合治理

右上图 /Top right

Cheng Hai (China): Speleothem and climate change

程海（中国）：洞穴沉积物与气候变化

右下图 /Bottom right

Zhang Yuanhai & Shen Lina (China): Technology and methodology of Tiankeng survey

张远海和沈利娜（中国）：天坑调查技术与方法

第四章　国际交流与培训

从左至右、从上至下依次为（按学员答辩顺序）

From left to right, top to bottom are (According to the training course schedule)

Matej Lipar (Slovenia): Speleothems from Postojna Cave, Slovenia; Boualla Othmane (Morocco): Collapse dolines in the Safi region (western Morocco): Mapping and geophysical approaches; Ross Dominic Darang Agot (the Philippines): Geohazard assessments in karst landscapes: The case of massive landslide incident in Naga city, Cebu Island, Philippines; Lovel Kukuljan (Slovenia): CO_2 dynamics and dissolutional processes in the karst vadose zone; Zhang Jian (China): Hydrological climate change revealed in the Villars Cave (SW-France) using geophysical, hydrological and isotopic methods; Beatriz Hadler Boggiani (Brazil): Iron ore caves in Brazil; Kyaw Zayya (Myamar): Mineral deposit and well-known caves in karst area, Myamar; Saeed Habeminezhad (Iran): Investigating the geochemistry of the speleothems in caves of NE Iran and paleoclimate studies

梅特椰·利帕（斯洛文尼亚）：斯洛文尼亚波斯托伊纳洞穴沉积物；保拉·奥斯曼（摩洛哥）：萨菲区（摩洛哥西部）塌陷漏斗填图及地球物理探测；罗丝·亚格特（菲律宾）：岩溶景观区地质灾害评估——以菲律宾宿务岛那牙市大型滑坡为例；诺娃·库库兰（斯洛文尼亚）：岩溶包气带CO_2动力学及溶解过程；张键（中国）：运用地球物理、水文学及同位素方法揭示Villars洞（法国西南部）的水文气候变化；贝报兹·波及尼（巴西）：巴西铁矿洞；玖赞亚（缅甸）：缅甸岩溶区矿产资源和著名洞穴；赛德·哈贝姆卡德（伊朗）：伊朗东北部洞穴沉积物地球化学调查及古气候研究

6. The 13th international training course on "Karst and Sustainable Development"
第十三届国际培训班——岩溶与可持续发展

The 13th International Training Course organized by IRCK and co-funded by the International Union of Geological Sciences (IUGS) was held from November 15 to 26, 2021. Since the outbreak of COVID-19, this is the second virtual training course held by IRCK. It serves as a platform for mutual learning and exchanges for karst scientists, natural resources and environmental researchers, and college students around the world. It is also one of the important activities of the International Year of Caves and Karst (IYCK), aiming to enable a better understanding and protection of karst by the public.

2021 年 11 月 15~26 日，岩溶中心组织实施了第十三届国际培训班，并由国际地质科学联合会联合赞助。本次培训班是受新冠疫情影响以来中心举办的第二次线上国际培训，为世界各地从事岩溶科学研究的学者、各国自然资源与环境领域从业者及各高校学生搭建了良好的交流学习平台。此次培训班同时是国际洞穴与岩溶年的重要活动之一，旨在使大众更好地理解、探索和保护岩溶。

The training course themed on "Karst and Sustainable Development" has attracted 41 trainees from 18 countries (Brazil, the Philippines, Iran, etc). To allow for attendance from various parts of the world, the training course was run in two phases according to different time zones. Fifteen experts from six countries like France, Italy, and Slovenia were invited as lecturers and evaluators. The lectures focused on four topics: karst carbon cycle and climate change, karst water resources and water security, karst ecological restoration and eco-industry, and karst landscape resources and sustainability. The effective organization of the training course ensures efficient communication between the trainees and lecturers. On November 26, the evaluation team for the training assessment selected 8 excellent trainees from the Republic of Congo, Pakistan, Tunisia, Mexico, Iran, the Philippines, and Brazil.

本届培训班的主题为"岩溶与可持续发展"，来自巴西、菲律宾、伊朗等 18 个国家的 41 名学员参加了此次培训。为增强培训效果，减小时差影响，本次培训班按时区分两期进行。此次培训班邀请到了来自法国、意大利、斯洛文尼亚等 6 国的 15 名专家担任讲师和评估专家，围绕岩溶碳循环和气候变化、岩溶水资源与水安全、岩溶生态修复和生态产业及岩溶景观资源与可持续利用4个方向线上授课，学员踊跃提问，互动交流热烈。11月26日，通过学员答辩和评委评议，来自刚果、巴基斯坦、突尼斯、墨西哥、伊朗、菲律宾和巴西的8位学员成为本届培训班优秀学员，获得了国际地质科学联合会奖励。

左上图 /Top left
Wang Hongmei (China): Subsurface biosphere in karst caves and their role in carbon sink
王红梅（中国）：岩溶洞穴地下生物圈及其在碳汇中的作用

右上图 /Top right
Pu Junbing (China): Carbon cycle in karst water
蒲俊兵（中国）：岩溶水体中的碳循环

左下图 /Bottom left
Bai Xiaoyong (China): Continental weathering carbon sink and global change
白晓永（中国）：大陆风化碳汇与全球变化

右下图 /Bottom right
Giuseppe Arduino (Italy): Central Italy Karst water resources and water security
朱赛佩·阿尔杜伊诺（意大利）：意大利中部岩溶水资源与水安全

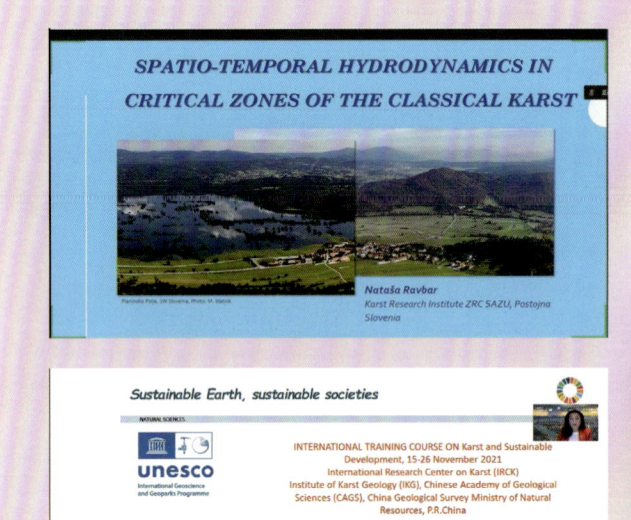

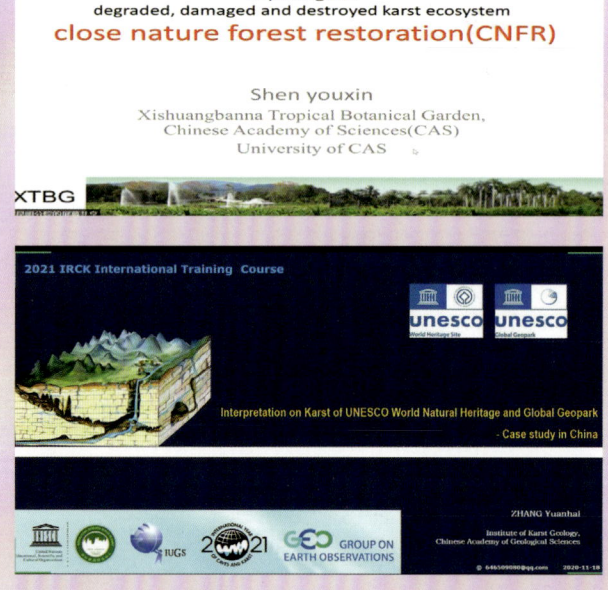

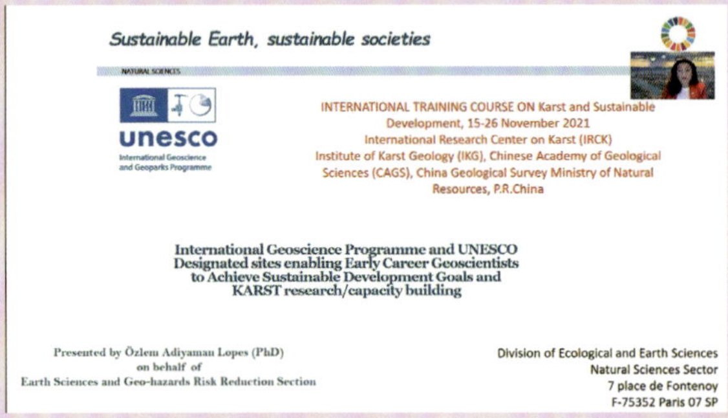

左上图 /Top left

Nataša Ravbar (Slovenia): Spatio-temporal hydrodynamics in critical zones of the classical karst

娜塔莎·阿尔巴（斯洛文尼亚）：经典岩溶区岩溶关键带水动力的时空变化

左下图 /Bottom left

Ozlem Adiyaman Lopes (UNESCO): International Geoscience Programme and UNESCO Designated sites enabling Early Career Geoscientists to Achieve Sustainable Development Goals and KARST research/capacity building

奥兹莱姆·洛佩斯（UNESCO）：国际地球科学计划和教科文相关机构助力地球科学工作者职业初期实现可持续发展目标及开展岩溶研究/能力建设

右上图 /Top right

Shen Youxin (China): Repairing the degraded, damaged and destroyed karst ecosystem—Close nature forest restoration (CNFR)

沈有信（中国）：修复退化、破坏的岩溶生态系统——封闭式天然林恢复

右下图 /Bottom right

Zhang Yuanghai (China): Interpretation on karst of UNESCO World Natural Heritage and Global Geopark—Case study in China

张远海（中国）：世界自然遗产和世界地质公园岩溶景观价值解译——中国案例

上图 /Top

Jiang Guanghui (China): Groundwater monitoring in the karstic China
姜光辉（中国）：中国岩溶区地下水监测

中图 /Middle

Zhou Jinxing (China): How to construct vegetation restoration and eco-economy development mode in karst rocky desertification area
周金星（中国）：岩溶石漠化区植被修复和生态经济开发模式

下图 /Bottom

Virtual trainee assessment
学员在线接受评估

4.2.2 Other international training courses
其他国际培训

1. Training Course on Great Mekong Subregion Geological Environment and Sustainable Development (Beijing)

大湄公河次区域地质环境与可持续发展培训研讨（北京）

From July 10 to 18, 2016, the Training Course on Great Mekong Subregion Geological Environment and Sustainable Development, which was hosted by the Ministry of Foreign Affairs of China, China Geological Survey, and organized by IRCK/IKG and the Institute of Hydrogeological and Environmental Geology (IHEG) under Chinese Academy of Geological Sciences, was held in Beijing. The training course invited 19 lecturers to give lectures or instructed field practice, and recruited 34 participants from 8 countries, including Cambodia, Indonesia, Laos, Malaysia, Myanmar, the Philippines, Thailand, and China. It focused on geological survey and cooperation of the partners along the Belt and Road, Maritime Silk Road and geological environment protection, water resources in karst area and karst collapse, as well as geological disasters and related database construction. During the training course, the Seminar on Sustainable Development of Groundwater and Geological Environment was also convened, the representatives of each country made country reports on hydrogeology and environmental geology of their own countries, and finally, all the participants reached a preliminary consensus that it is necessary for China and ASEAN member states to carry out further cooperation on hydrogeology and environmental geology in future.

2016 年 7 月 10~18 日，由外交部与中国地质调查局主办，中国地质科学院岩溶地质研究所 / 水文地质环境地质研究所、联合国教科文组织国际岩溶研究中心承办的"大湄公河次区域地质环境与可持续发展"培训研讨班在北京召开。本次培训班邀请了 19 名教员授课或指导野外实习，招收了来自柬埔寨、印度尼西亚、老挝、马来西亚、缅甸、菲律宾、泰国和中国 8 个国家的 34 名学员。培训围绕"一带一路"地质调查与合作、"海上丝绸之路"与地质环境保护、岩溶区水资源及岩溶塌陷、地质灾害及数据库建设等方面展开授课和讨论。其间,召开了"东盟地下水与地质环境可持续发展国际合作交流"研讨会,各国学员对本国的水文地质环境地质研究与调查情况进行了介绍,各方对进一步开展中国–东盟水文地质环境地质合作达成了初步共识。

The training course helped China to share the experience on groundwater utilization and geological environment protection to ASEAN countries, guiding them to develop groundwater resources reasonably and protect geological environment effectively. It promoted further cooperation on the regional geological environment of the Great Mekong Subregion, fostering a joint, win-win, and harmonious development for the Maritime Silk Road.

此次培训面向东盟国家,分享我国地下水与地质环境保护经验,引导其合理开发利用地下水资源,有效开展地质环境保护,进一步推动大湄公河次区域地质环境保护合作,为促进海上丝绸之路共建、共赢、和谐发展做出贡献。

上图 /Top

Choup Sokuntheara (Cambodia): Current status of groundwater and geology in Cambodia
乔普·素坤塔拉(柬埔寨):柬埔寨地下水现状及地质概况

下图 /Bottom

Arief Kurniawan (Indonesia): Country Report of Indonesia
阿里夫·昆尼万(印度尼西亚):印度尼西亚国家报告

Chapter 4 International Exchange and Training

从左至右、从上至下依次为（按授课顺序）
From left to right, top to bottom are (According to the schedule)
Ounakone Xayviliay (Laos): Lao PDR sustainable use of ground water; Mohd Khairul Nizar Shamsuddn (Malaysia): Malaysia geology and hydrogeology overview; Kyaw Thu Aung & Nyein Ar Kar Zaw (Myanmar): Groundwater survey and utilization & geo-environment sustainable development of Great Mekong Subregion; Marnette B. Puthenpurekal & Ma. Teresa I. Coching (the Philippines): Geo-environmental protection, geology, and groundwater resources of the Philippines; Mahippong Worakul (Thailand): Thailand's national report on geological environmental issues: Groundwater resources and geo-hazard management; Cheng Yanpei (China): Hydrogeological investigation, map compiling, and monitoring in Mekong River Basin; Zhang Fawang (China): Initiative of China-ASEAN cooperation on trans-boundary mapping of hydrogeology and environmental geology

奥拉坤·夏维里（老挝）：老挝地下水资源可持续利用；莫德·沙马赤丹（马来西亚）：马来西亚地质和水文地质概况；玖图昂和内亚卡赞（缅甸）：大湄公河次区域地下水调查与利用及地质环境可持续开发；玛尼特·普丹拉克尔和马特蕾沙·克钦（菲律宾）：菲律宾地质环境保护、地质及地下水资源；马海鹏·沃拉库（泰国）：泰国地质环境问题报告——地下水资源和地质灾害治理；程彦培（中国）：湄公河流域水文地质调查、编图及监测；张发旺（中国）：中国-东盟关于跨界水文地质环境地质编图合作倡议

The Second 6 Years of IRCK

Group photo of the training course
培训班合影

2. Training Course on "Sustainable Development of Groundwater and Geological Environment in the Great Mekong Subregion" (Bangkok)
大湄公河次区域地下水与地质环境可持续发展培训研讨（曼谷）

On August 22-28, 2016, the Training Course on "Sustainable Development of Groundwater and Geological Environment in the Great Mekong Subregion", jointly organized by IRCK/IKG, the Department of Groundwater Resources of Thailand (DGR), and the Department of Mineral Resources of Thailand (DMR) was held in Bangkok. The training course lasted for 7 days, with 43 trainees from 8 countries, including China, Thailand, Laos, Cambodia, Myanmar, Vietnam, Indonesia, and Malaysia. The lectures focused on the geological environmental problems and sustainable development in Thailand, including the geological and environmental problems along the China-Thailand railway, seawater intrusion, land subsidence, etc. Lecturers from IRCK introduced hydrogeology and environmental geology, database construction, groundwater development and evaluation, and so on. Moreover, the trainees visited the National Geological Museum of Thailand and the Chedi Hoi Temple, having a better understanding of the geological research status of Thailand and their culture. In addition, IRCK delegation had a meeting with DMR, DGR, and CCOP respectively for further cooperation with the minutes signed by relevant parties.

2016年8月22~28日，中心联合泰国地下水资源局、矿产资源局举办的"大湄公河次区域地下水与地质环境可持续发展"培训研讨班在泰国曼谷举行。本次培训班为期7天，共有来自中国、泰国、老挝、柬埔寨、缅甸、越南、印度尼西亚和马来西亚等8个国家的43名学员参加了培训。培训班围绕泰国地质环境问题与可持续发展展开研讨，包括中泰铁路沿线地质环境问题、海水入侵、地面沉降等，中心教员还围绕水文地质与环境地质、数据库建设、地下水开发与评价等进行授课。最后，学员参观了泰国国家地质博物馆及贝壳塔寺，了解了泰国的地质研究概况及典型特色文化。其间，中心代表团分别与泰国地下水资源局、矿产资源局及CCOP进行了会谈，针对下一步具体合作签署了会谈纪要。

The training course deepened the cooperation on hydrogeology and environmental geology between China and Thailand, consolidated current cooperation patterns, and helped all the stakeholders to reach a consensus on promoting of groundwater resources and geological environment. The training course made some contribution to the construction of Maritime Silk Road and the common development of Lancang-Mekong community.

本次培训班深化了中泰水文地质、环境地质合作，巩固了中泰双方现有的地下水与地质环境合作模式，初步达成了推进地下水与地质环境可持续发展的广泛共识，为推进"海上丝绸之路"建设、实现澜湄命运共同体发展愿景起到了一定促进作用。

Group photo of the training course
培训班合影

Name: Fengshan National Geopark
Location: Guangxi
Inscribed as a National Geopark in 2005
Summary: Fengshan National Geopark is known for its groupings of large caves, with huge speleothems, windows of ground rivers, and the world's highest underground sinkhole valley. Fengshan has the second largest natural bridge in China and unique karst springs.

名称：广西凤山国家地质公园
所在地点：广西
列入国家地质公园时间：2005年
概述：作为国家地质公园，凤山是我国大型洞穴分布最为密集的地区，拥有世界大型石笋群、世界天窗群、世界最高的地下溶洞峡谷、中国跨度第二的天生桥、千古之谜鸳鸯泉等独特的地质遗迹景观。

Chapter 5

Scientific Popularization and Consultation

第五章　科学普及与咨询

Chapter 5 Scientific Popularization and Consultation

From 2016 to 2021, IRCK carried out serial popular science activities mainly in the China Museum of Karst Geology (Karst Museum) and IRCK's field experimental sites, edited popular science products, and promoted karst science popularization and consultation. IRCK hopes to raise public's awareness which is a crucial foundation for environmental protection and resources sustainable use of karst areas.

2016~2021年，中心以中国岩溶地质馆、野外试验基地为主要活动场地，积极开展科普活动，编制科普产品，推进岩溶科学普及与咨询服务。此举旨在提高公众意识，从而为岩溶区环境保护和资源可持续开发利用奠定基础。

5.1 Carry Out Thematic Activities and Enable Effective Science Popularization Effects
推进主题科普活动，深化日常科普成效

From 2016 to 2021, IRCK carried out more than 200 popular science activities with more than 6,500 participants involved. Especially, on/in Earth Day, Science and Technology Week and National Land Day, IRCK organized important events assembling Karst Museum visits, popular science in schools, virtual competitions, and other activities to publicize karst science, which aroused teenagers' strong interest.

2016~2021年，中心共开展科普活动200余次，参与人数6500余人。中心围绕"世界地球日""科技活动周""全国土地日"开展主题科普活动，将博物馆参观、科普进校园、在线竞赛有机结合，积极宣传岩溶科学知识，激发青少年学习兴趣。

5.1.1 Thematic activities
主题科普活动

The 47th Earth Day (2016)
第47个世界地球日（2016年）

The 48th Earth Day (2017)
第 48 个世界地球日（2017 年）

The 47th and 48th Earth Day (2016, 2017) — Using Resources Economically and Intensively, Advocating Living Greenly and Simply: More than a hundred visitors from universities, research institutions, and enterprises in Guilin took part in the Earth Day activities, which promoted a conception that "Advocating green travel, using karst resources greenly, and protecting our environment by our actions".

第47、48个世界地球日（2016年、2017年）——节约集约利用资源，倡导绿色简约生活：来自桂林市高校、科研院所和企业的百余人参与了主题科普活动，活动推广了"倡导绿色出行，节约利用岩溶资源，以我们自身的实际行动保护我们自己所生存的环境"的理念。

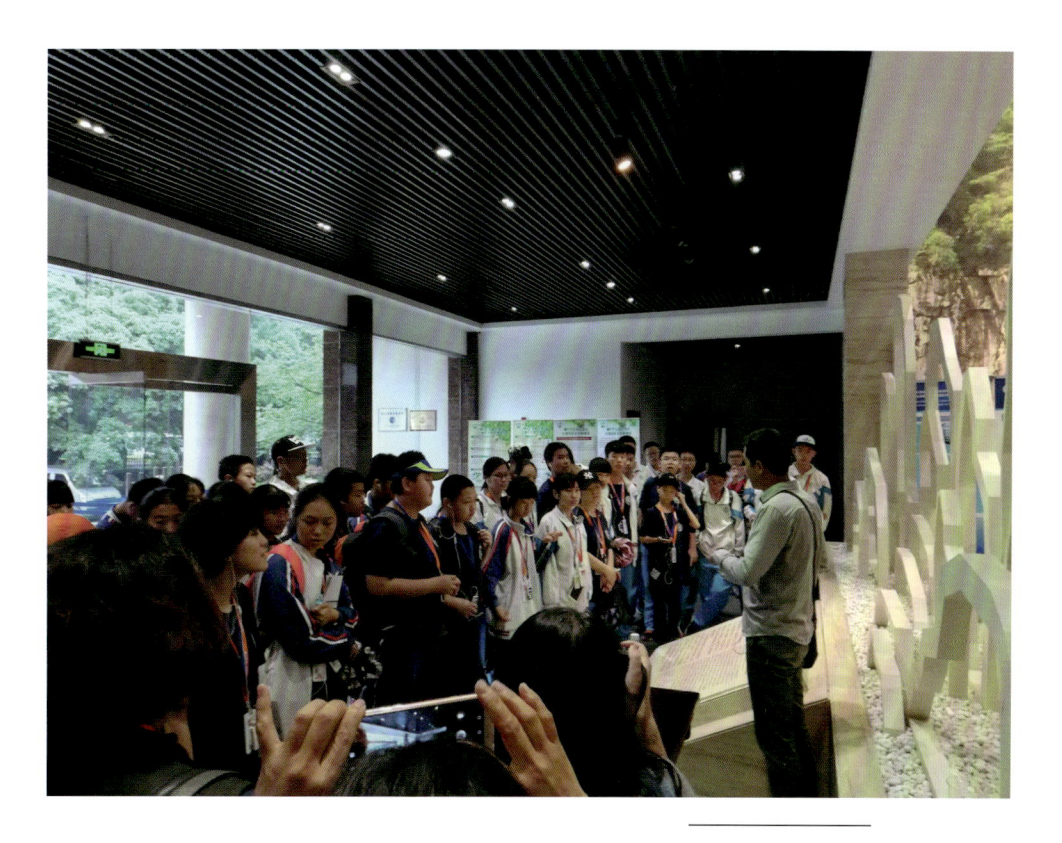

The 49th Earth Day (2018)
第 49 个世界地球日（2018 年）

The 49th Earth Day (2018)— Cherishing Natural Resources, Take Good Care of Our Beautiful Land: IRCK organized serial activities including science popularization in schools, karst science camp, and "Go to Jinfo Mountain", also received visitors from Beijing Yinghuayuan Experimental School and Shaoyang University in Hunan to visit Karst Museum.

第 49 个世界地球日（2018 年）——珍惜自然资源，呵护美丽国土：中心开展了科普讲座进校园活动、岩溶科学营活动和走进金佛山开展"世界地球日"科普活动，组织北京樱花园实验学校、湖南邵阳学院的师生参观中国岩溶地质馆。

The flag-giving ceremony for the first "Geologist Li Siguang Squadron" in Guilin
桂林首支"李四光中队"授旗仪式

The 50th Earth Day (2019)— Cherishing Beautiful Earth, Guarding Natural Resources: IRCK organized Karst Museum tour, popular science posters exhibition, and special lectures in schools, in order to publicize earth science and help the public understand more about the current status of resources and the environment in China.

第50个世界地球日（2019年）——珍爱美丽地球，守护自然资源：中心组织了参观中国岩溶地质馆、参观专题科普展板、开展入校科普讲座活动，宣传地球科学知识，使人们深入了解我国资源和环境现状。

Prof. Jiang Zhongcheng, the Governing Board member, granted the Board of Teenager Science Popularization Education Base to Yucai Primary School
中心理事蒋忠诚研究员为育才小学授牌 —— 青少年科普教育基地

The 51st Earth Day (2020)— Cherishing Earth, Harmonious Coexistence between Human and Nature: Influenced by COVID-19, the activity was organized both virtually and on spot. Academician Yuan Daoxian (top), the Governing Board member, also the director of the Academic Committee, together with Prof. Chen Weihai, scientist of IRCK, also the chief scientist of National Key Research & Development Program, made virtual popular science lectures (The first "Geologist Li Siguang Squadron" listened to the virtual lectures in classroom (opposite page)); meanwhile, IRCK held a virtual competition for popular science. In addition, popular science articles about karst groundwater, geological heritages, and karst collapse were published through websites and

WeChat official account, enabling the public to raise their awareness through a virtual learning plat form at home.

  第 51 个世界地球日（2020 年）   珍爱地球，人与自然和谐共生：受新冠疫情影响，本次系列科普活动以线上为主、线下为辅形式开展。本次主题科普活动邀请了中心理事、中国科学院院士、学术委员会主任袁道先先生（对页图），以及中心副总工程师、国家重点研发计划项目首席科学家陈伟海等科普专家，通过网络课堂的形式开展科普讲座（桂林市首支"李四光中队"参加网络科普知识讲座（上图））；同步开展线上科普知识竞赛。此外，通过网站和微信公众号刊登岩溶地下水、地质遗迹、岩溶塌陷等方面的科普文章，让社会公众足不出户即可了解岩溶科普知识，提升公众意识。

IRCK provided popular science posters in Guilinyang Middle School of Haikou, Hainan
中心在海口市桂林洋中学开展科普展览活动

The 52nd Earth Day (2021)— Cherishing Earth, Harmonious Coexistence between Human and Nature: IRCK organized various interesting science activities like museum visit, lectures, virtual competition, and short video competition to publicize karst science.

IRCK provided popular science lectures in Guilinyang Middle School of Haikou, Hainan
中心在海口市桂林洋中学开展科普讲座活动

第52个世界地球日（2021年）——珍爱地球，人与自然和谐共生：通过参观、讲座、线上竞赛、短视频竞赛等形式多样、妙趣横生的活动，针对在校学生开展科普。

5.1.2 Science and Technology Week
科技活动周

In 2017, taking "Science and Technology Promote Our Nation, Innovation Enable Our Success" as the theme, IRCK attended Guangxi Innovation-Driven Development Achievement Exhibition in Science and Technology Week.

2017年，中心以"科技强国 创新圆满"为主题，参加了科技活动周广西创新驱动发展成果展。

In 2019, taking "Science and Technology Promote Our Nation, Science Popularization Benefits Our People" as the theme, IRCK participated in the 28th Science and Technology Week·Guangxi Innovation-Driven Development Achievement Exhibition, with important international cooperation results, geoheritage investigation, as well as UGGps and World Heritage application exhibited, as well as the geophysical survey results in deep earth, attracting great attention from the public.

2019年中心以"科技强国 科普惠民"为主题，参加了第二十八届广西科技活动周·广西创新驱动发展成果展活动，重点展出了一系列国际合作成果以及地质遗迹调查与地质公园、世界遗产申报和深部地球探测等社会服务项目成果，得到广大民众的高度关注。

In 2020, IRCK organized serial science popularization activities themed on "Fighting Against the Pandemic by Science and Technology, Strengthening Our National by Innovation". IRCK made advertisements through Guilin Science and Technology Bureau, and local media like *Guilin Daily* and *Guilin Evening News*, attracting over 300 visitors to Karst Museum; moreover, IRCK organized a teenagers' painting competition themeing on "Protecting Lijiang River Scientifically" with Guilin Science and Technology Bureau jointly. In 2021, IRCK organized the 2nd painting competition for Protecting Lijiang River Scientifically, with 17 paintings awarded.

2020年科技活动周，中心组织主题为"科技战疫 创新强国"系列科普活动。通过桂林市科技局以及《桂林日报》《桂林晚报》等本地媒体对中国岩溶地质馆场馆开放周的宣传，参观人数超过300人，联合桂林市科技局合作组织"科学保护漓江"主题青少年科普绘画竞赛。2021年科技活动周期间，中心举办第二届"科学保护漓江"小学生公益绘画大赛，共选出17幅获奖作品。

The awarded painting "Earth Protection" in 2020
2020年获奖作品"保护地球"

5.1.3 Other activities
其他科普活动

桂林是岩溶研究的圣地，其峰林地形以"中国式岩溶"而闻名，其地表和地下、宏观和微观、溶蚀和沉积形态齐全，而且地表宏观岩溶形态规模大，正负岩溶形态反差强烈，是进行岩溶形态组合研究、岩溶地貌演化和岩溶动力系统研究的理想场所。中国科学院袁道先院士（前排右5）几乎每年都带领中心科技人员与中国地质科学院、西南大学、桂林理工大学、中国地质大学（武汉）的研究生进行岩溶地貌、岩溶形态野外实习教学。

Guilin, as the holy land for karst research, is well-known because of its "Chinese Style Karst"—the peak forest. Guilin has nearly all kinds of surface and subsurface, macroscopic and microscopic, dissolved and sedimentary karst morphology. Moreover, the surface macroscopic karst morphology in Guilin is grand, featured by contrasting positive and negative morphology. Guilin is an ideal place to research the combination of karst morphology, karst landform evolution, and karst dynamic systems. Academician Yuan Daoxian (row 1, the fifth from the right) led the young scientists from research institutions like Chinese Academy of Geological Sciences, Southwest University, Guilin University of Technology, and China University of Geosciences (Wuhan) to have a field study for karst landforms and morphology nearly every year.

In 2018, the National Geographic Society of the U.S. organized 27 scholars from different fields to visit IRCK. Dr. Cao Jianhua, the executive deputy director of IRCK (top, the second from left; bottom, the second from left) introduced the latest achievements of IRCK and the basic knowledge of karst science to the visitors in detail. Meanwhile, he led them to visit the Karst Museum (top) and the Maocun Field Experimental Site (bottom), and had a heated discussion about the groundwater migration rules and karst landform formation.

2018年，美国国家地理学会组织美国各领域的27名专家教授访问联合国教科文组织国际岩溶研究中心。中心常务副主任曹建华研究员（上图，前排左2；下图前排左2）介绍了岩溶中心近年成果，为来访人员详细介绍了岩溶科学基础知识，此外，他还带领来访人员参观中国岩溶地质馆（上图），考察了毛村野外试验场（下图），围绕岩溶区地下水的运移规律及岩溶地貌等与来访人员展开了热烈讨论。

5.2 Refining Popular Science Achievements Enabled Edutainment Popular Science Products
提炼科普成果，编制寓教于乐的科普产品

From 2016 to 2021, IRCK published 51 popular science readings, produced 3 popular science videos, and developed 2 sets of teaching materials for studying tour. IRCK carried out scientific popularization and dissemination work about karst groundwater, karst landscape resources, karst geological disasters in simple but vivid words.

2016~2021 年，中心出版科普读物 51 种，科普视频 3 个，开发研学教材 2 套。中心分别围绕岩溶地下水、岩溶景观资源、岩溶地质灾害等进行深入浅出的科学普及工作。

5.2.1 Popular science readings
科普读物

Karst Geology and Ecosystem — Karst Science Popularization in China: Working with the Chinese National Committee for Man and the Biosphere Programme, IRCK edited the book *Karst Geology and Ecosystem—Karst Science Popularization in China* (in both English and Chinese), which included the causes of karst and rocky desertification, karst and carbon cycle, karst and ecosystem, and karst as a witness of the evolution of the earth. The brochure introduced the phenomenon and theory by elegant photos and graphic words.

《岩溶与生态——岩溶科普在中国》：中心联合中华人民共和国人与生物圈国家委员会，编制了《岩溶与生态——岩溶科普在中国》（中英文版）一书，该书介绍了岩溶与石漠化成因、岩溶与碳循环、岩溶与生态系统及岩溶见证地球演变等方面的内容。这本小册子以精美照片，辅以生动语言，深入浅出地阐述了岩溶科学的现象及原理。

岩溶与生态
岩溶科普在中国

KARST GEOLOGY
AND ECOSYSTEM
KARST SCIENCE
POPULARIZATION IN CHINA

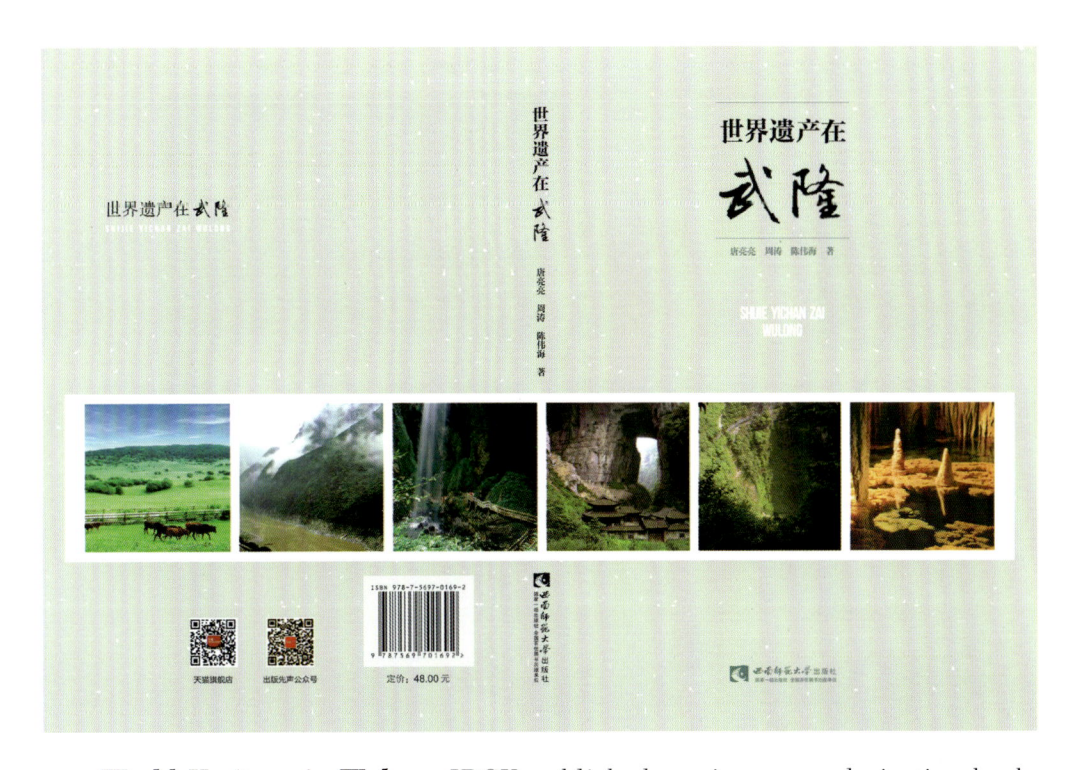

World Heritage in Wulong: IRCK published a science popularization book *World Heritage in Wulong*, explaining the formation processes, features, and "great significance and universal value" of the World Natural Heritage Sites-South China Karst, especially the Wulong Karst, in plain language. The book introduced how Chinese natural heritage sites, including the Wulong Karst, were "inscribed" on the "World Heritage List" and analyzed the current management and development model of world heritage sites over the world. The book serves as a reminder for the public to protect world heritage sites, and hopes to ensure that the future generations could enjoy the rewards from the nature.

《世界遗产在武隆》：该书以通俗易懂的语言阐述了中国南方喀斯特世界自然遗产，特别是武隆喀斯特世界自然遗产的形成过程、景观特征及"突出意义和普遍价值"，介绍了武隆喀斯特的"申遗"之路，以及中国的漫漫"世遗"之路，分析了当前国内外世界遗产的管理和开发模式，提醒人们要保护好世界遗产，让全人类和子孙后代永享大自然的恩泽。

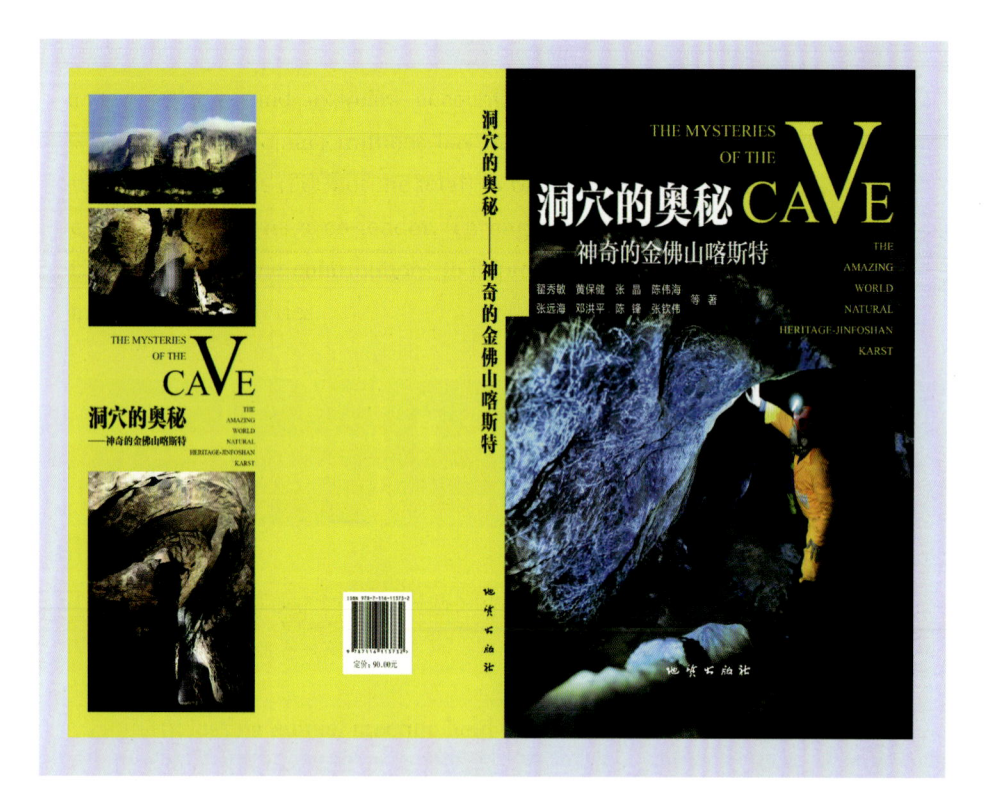

The Mysteries of the Cave—Amazing Jinfo Mountain Karst: This book explains the caves in Jinfo Mountain in an accessible manner. It introduced distribution, developing characteristics, scales, physical accumulations, speleothems, long history for human activities (e.g. boiling nitrate for gunpowder) in Jinfo Cave, the hidden scientific mysteries, and the evolution processes of caves in Jinfo Mountain, which condensed the geological and scientific values of Jinfo Mountain Karst.

《洞穴的奥秘——神奇的金佛山喀斯特》：该书浅显易懂地介绍了金佛山洞穴分布及发育特征、洞穴空间规模、洞穴物理堆积及次生化学沉积物、金佛洞悠远的利用历史（熬硝制造火药）、洞穴中隐藏的科学奥秘、洞穴发育演化过程及阶段，凝练了金佛山遗产地的地学价值和科普价值。

The Source of the Pearl River — Magical Karst: The book summarizes multi-year investigation and research results of karst underground rivers, and popularizes the knowledge about underground rivers to the public like the concept, source, distribution, searching methods, potential hazards, as well as development, utilization, and protection.

《珠江源——神奇的岩溶》：该书梳理多年来岩溶地下河调查与综合研究成果，向大众普及地下河概念、来源、分布、寻找方法、潜在危害、开发利用与保护等内容。

Popular Science Atlas for Sr-rich Mineral Water in Xintian County: The atlas included 11 maps like the Sr-rich Mineral Water Fields Distribution Map, I-rich Mineral Water Distribution Map, Sr-rich Mineral Water Fields Development, Utilization, and Planning Map, Water Resources Protection Zoning Map, Groundwater Movement Sketch Map, etc. The atlas was handed over to the local government, which may serve for reasonable utilization of water resources.

《新田县大型富锶矿泉科普小册》：该图册包括富锶矿泉水田分布图、富碘矿泉水分布图、富锶矿泉水田开发利用规划图、水资源保护区划图、地下水运动示意图等11幅图，并移交当地政府，以利于合理开发利用水资源。

Re-exploring the Tiankeng Group in Hanzhong: The *Chinese Newspaper of Land and Resources* used a full page to introduce the discovery, numbers, and significance of the tiankeng group in Hanzhong, as well as the achievements of the fourth China-Czech joint expedition after re-exploring the Hanzhong Tiankeng Group. Hanzhong Tiankeng group could be considered as an evidence to move the current boundary of humid tropic-subtropic karst landform northwards, which has a great significance to the analysis and research of ancient geographic environment and climate change.

《再探汉中天坑群》：《中国国土资源报》地质调查版块整版刊登《再探汉中天坑群》，介绍了汉中天坑群的发现过程、数量、意义，以及中捷第四次联合科考"再探汉中天坑群"取得的成果。汉中天坑群的发现使得我国湿润热带－亚热带岩溶地貌区界线显著北移，对古地理环境及气候变化的分析研究具有重要意义。

The Missing CO₂: By cooperation with Guangxi Institute of Botany under Chinese Academy of Sciences, and China Central Television, IRCK has made a film about *the Missing CO₂*. The film presents the research progress of Maocun Field Experimental Site, with a large amount of scientific data, charts, three dimensional animations, etc. It explained the atmospheric CO_2 change and the causes vividly, serving as an effective product to popularize karst science.

《失踪的 CO_2》：中心联合中国科学院广西植物研究所、中央电视台等拍摄完成《失踪的 CO_2》。该片展示了毛村野外实验基地的研究进展，结合大量科学研究数据、图表、三维模拟动画等，从岩溶碳循环及碳汇的角度，形象地讲解了大气中 CO_2 含量变化及其原因，是良好的科普宣传素材。

Xingyi National Geopark "Three Character Classic" Poem: This poem introduces the geological evolution story of Guizhou Xingyi National Geopark, which was played in Xingyi TV Station, Xingyi National Geopark Museum, Xingyi Geopark "Fossil Science Popularization Week", and the Fifth Asia Pacific Geopark Network Conference (Zhijin County, Guizhou) with positive feedback from all the audiences.

《兴义国家地质公园科普三字经》：该片介绍贵州兴义国家地质公园的地质演化故事，在兴义市电视台、兴义国家地质公园博物馆、兴义地质公园"化石科普周"、第五届亚太地质公园会议（贵州织金县）展播，反响良好。

Karst Science Popularization Series: Caves: Taking lollipops as an example, the film explained the formation of caves from the limestone dissolution, collapse, and transport by the underground rivers to the formation of various beautiful speleothems by the dripping water. It advocates the public to cherish the gifts endowed by nature when they are enjoying the beautiful scenery of the caves.

《岩溶科普系列：洞穴》：以棒棒糖的溶解为例，该片从溶洞的溶蚀，讲述石灰岩溶解，到溶洞崩塌和地下河水搬运的过程，最后讲述溶洞滴水形成的千姿百态的钟乳石，号召大家在享受洞内美景的同时，珍惜自然的馈赠。

5.2.2 Studying tour products
研学产品

Mysterious World of Caves **(A teaching reading)**: This reading introduces the origin and distribution of karst, as well as the definition, classification, formation, evolution, sediments, and creatures of caves. It is an accessible but informative reading.

《神秘的洞穴世界》研学教材：该教材介绍了岩溶的由来及分布、洞穴的定义和分类、洞穴的形成与演化、洞穴沉积物、洞穴生物等内容，内容丰富，浅显易懂。

Chapter 5 Scientific Popularization and Consultation

中国南方喀斯特-桂林山水甲天下

中国地质科学院岩溶地质研究所

South China Karst-Guilin Landscape, One of the Best in the World (**A teaching reading**): This reading introduces typical karst morphology in different climatic zones, and compares them with the special cone karst in Guilin. The reading was evaluated as one of the excellent courses for studying tours by the Geological Society of China.

《中国南方喀斯特－桂林山水甲天下》研学教材：该教材介绍了不同气候带典型的岩溶形态，并与桂林独特的锥状喀斯特进行对比。其研学课程成功申报中国地质学会精品地学科普研学课程。

Name: Leye-Fengshan UNESCO Global Geopark
Location: Guangxi
Listed as UNESCO Global Geopark in 2010
Summary: Leye-Fengshan UNESCO Global Geopark is located in western, Guangxi, with a total area of 930 km². The park has the largest dolinie group, the most concentrated group of cave hall, karst window group, the natural bridge with largest span length, typical cave sediments, the most completed fossil. It has important scientific and aesthetic value.

名称：广西乐业—凤山世界地质公园
所在地点：广西壮族自治区百色市凤山县
列入世界地质公园时间：2010年
简述：中国乐业—凤山世界地质公园位于广西西北部云贵高原向广西盆地过渡的斜坡地带，总面积930km²。公园内拥有全球最大的天坑群，最集中分布的洞穴大厅群、天窗群，最大跨度的天生桥，典型洞穴沉积物，最完整的早期大熊猫小种头骨化石以及独特天坑生态环境保留的动植物多样性，具有重要的科学研究意义以及极高的美学观赏价值。

Chapter 6

Summary

第六章 总结

Chapter 6 Summary

This chapter will introduce the major achievements of IRCK according to the contributions to different stakeholders, it is hoped that this chapter may provide different interpretations from different perspectives to help the readers to know more about IRCK, so as to help them understand IRCK's original aspirations and missions.

本章将按照为不同利益相关方所做的贡献，分别介绍中心主要成果，为读者进一步了解中心历程提供不同视角的解读，从而帮助读者理解中心努力奋斗的初心与使命。

6.1 Contributions to Realize the Objectives and Functions of the Renewal Agreement
为实现二期协定目标与职能所做的贡献

IRCK has realized the objectives and functions set in the renewal agreement (see Section 1.1 for details), played an active role as an international platform, and fulfilled the commitments in the agreement earnestly.

中心较好地完成了第二期协定中设定的目标与职能(详见1.1节),充分发挥了国际平台功能,切实履行了协定中的各项承诺。

1. Scientific research
科学研究

IRCK launched the International Big Scientific Plan on "Global Karst", and proposed long-term practical and effective solutions to resources and environmental problems in karst areas through bilateral and multilateral international cooperation in such fields as karst and climate change, karst water resources and water security, karst ecosystem and eco-industry, karst geoheritage diversity protection and sustainable utilization, and karst geohazards prevention and early warning. A total of 39 experts from 22 countries and 3 international organizations signed support letters to show their support for Global Karst. A total of 25 countries cooperated with IRCK to implement Global Karst, realizing corresponding objectives and functions in the Renewal Agreement.

中心启动了"全球岩溶动力系统资源环境效应"国际大科学计划,通过开展岩溶与气候变化、岩溶水资源与水安全、岩溶生态系统与生态产业、岩溶地质遗迹多样性保护与可持续利用、岩溶地质灾害防治与预警等领域双多边国际合作,切实为岩溶区资源环境问题提出了长效解决方案。共计有来自22个国家和3个国际组织的39名中外专家签署了支持函以示对该计划的支持,共计有25个国家参与大科学计划项目合作,较好地实现了中心协定中的对应目标与职能。

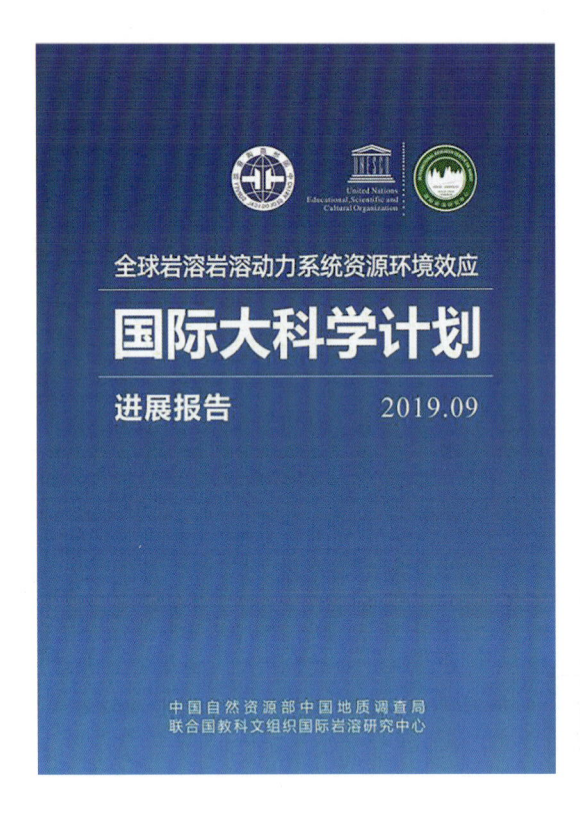

Progress Report of Global Karst
《国际大科学计划进展报告》

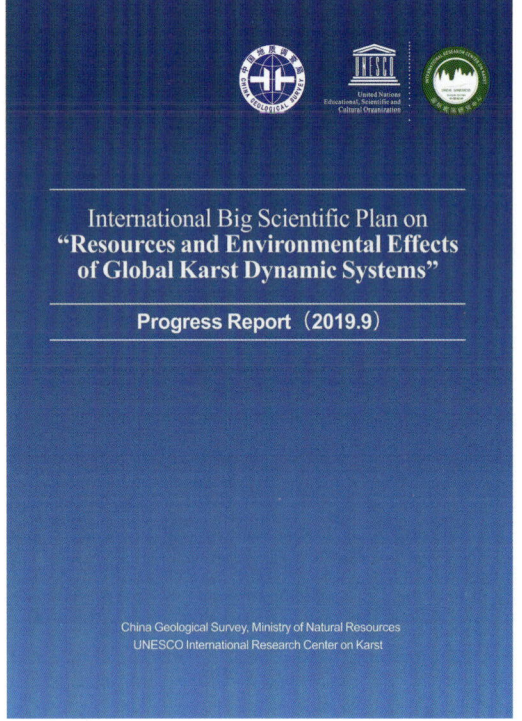

2. Publishing activities
出版活动

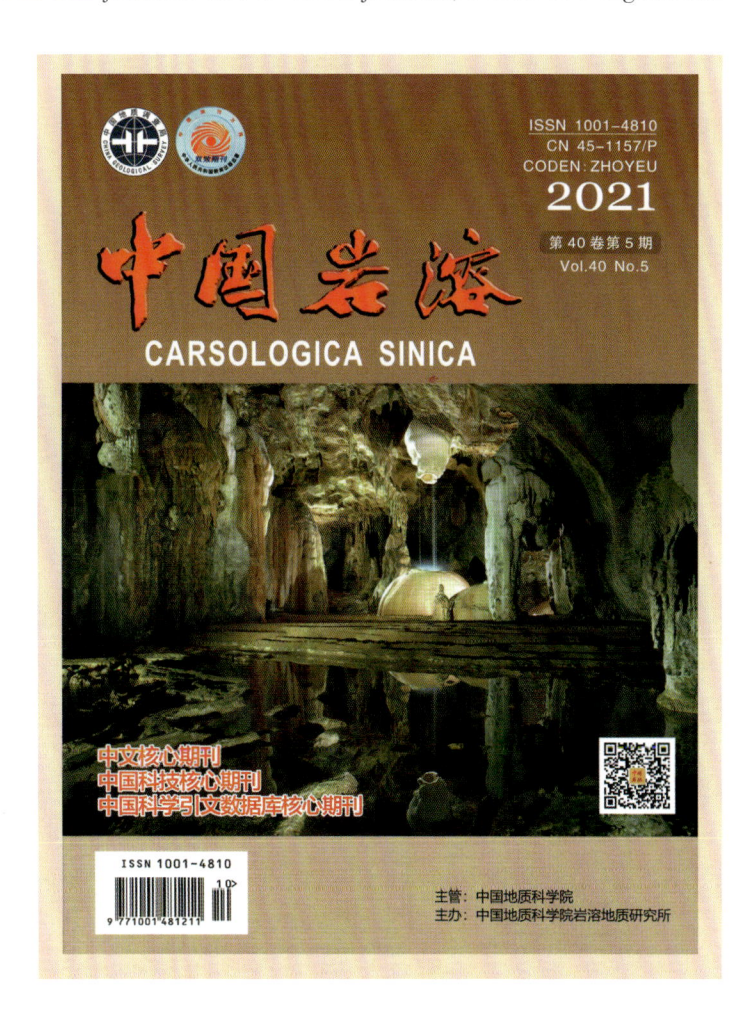

IRCK has published (generated) 18 monographs in total, including 16 academic monographs and 2 popular science monographs; 700 articles in Chinese core or above journals, including 157 in SCI journals and 69 in EI journals; a total of 7 significant atlases; and sponsored a journal titled as *Carsologica Sinica*. Through publishing activities, IRCK has publicized the scientific research achievements, promoted the development of karst dynamics, and realized the basic demand that scientific research should serve for the society.

中心共计出版专著18本，其中学术专著16本，科普专著2本；发表中文核心以上文章近700篇，其中SCI检索论文157篇，EI检索论文69篇；编制重要图集7份；主办《中国岩溶》期刊1份。中心通过出版活动，宣传了科研成果，促进了岩溶动力学发展，实现了科研为社会服务的根本需求。

3. International cooperation
国际合作

The cooperation among IRCK and international organizations is more extensive and further. Firstly, IRCK successfully set up the Karst Technical Committee under the International Organization for Standardization (ISO/TC 319), and successfully joined the International Union of Geological Sciences (IUGS) and the Group on Earth Observation (GEO). The success indicates a special international platform for karst standardization, a high recognition of karst discipline development from the authoritative geological organization, and a guarantee for the resources on a global-scale karst observation from a professional and influential international organization. Second, IRCK applied an International Geoscience Program (IGCP 661) successfully, and supported the successful application of Xiangxi UNESCO Global Geopark, realizing a good cooperation with the International Geoscience and Geopark Program (IGGP); cooperated with the Chinese National Committee for Man and the Biosphere (MAB) to make a popular science reading; and took part in the work of the World Karst Aquifer Map (WOKAM) project sponsored by International

Hydrological Program (IHP). Third, during the tenures of Executive Deputy Director, Secretary General, and Academic Committee Members, they were elected to hold important positions in international organizations such as the Karst Commission of the International Geographical Union (IGU-KC), the Karst Commission of the International Association of Hydrogeologists (IAH-KC) and the International Union of Speleology (UIS), enhancing the international influence of IRCK further. Fourth, IRCK organized or co-organized 18 important international or domestic conferences, and participated in 29 international academic conferences related to karst, which promoted international exchanges for karst science and helped IRCK to fully play as a platform.

中心与国际组织的合作范围更加广泛、合作程度更加深入。其一，中心成功申报国际标准化组织岩溶技术委员会（ISO/TC 319），成功加入国际地质科学联合会（IUGS）和地球观测组织（GEO），岩溶标准化建设拥有了专项平台，岩溶学科发展获得了地学领域权威国际组织的高度认同，全球尺度岩溶观测获得了国际专业组织的资源保障。其二，中心成功申报国际地球科学计划项目一项（IGCP 661），支撑湖南湘西等成功申报世界地质公园，实现了与国际地球科学与地质公园计划（IGGP）的良好合作；联合中国人与生物圈（MAB）国家委员会合作编制科普读物一册；参与国际水文计划（IHP）世界岩溶含水层图项目一项（WOKAM）。其三，中心常务副主任、秘书长、学术委员在任期间，当选为国际地理联合会岩溶专业委员会（IGU-KC）、国际水文地质学家协会岩溶专业委员会（IAH-KC）、国际洞穴联合会（UIS）等国际组织的重要职务，进一步提升了中心的国际影响力。其四，中心举办、协办重要国际国内会议18次，参与岩溶相关国际学术会议29次，促进了岩溶领域国际交流，中心平台职能得以充分发挥。

4. Advisory activities and technical information
咨询服务、技术信息

IRCK researchers provided technical information and advisory service for local governments actively through more than 200 projects, focusing on karst ecological environment investigation and evaluation, karst carbon cycle and addressing the climate change, water resources investigation, development and protection, karst geological disaster prevention and control, and karst landscape resources development and utilization, which acted as scientific and technological support for harmonious development between environment and socio-economy of locals, translated scientific achievements into productivity effectively.

中心科技人员积极为各地方政府提供岩溶生态环境调查评价、岩溶碳循环与应对气候变化、水资源调查开发与保护、岩溶地质灾害防治、岩溶景观资源开发利用等各类技术信息及咨询服务共计200余项，为地方政府的生态环境与经济社会和谐发展提供了科技支撑，切实将科学成果转化为了实际生产力。

5. Training
培训

IRCK has organized 8 training courses for global karst-related postgraduate students, scientists, technicians, and managers during the second six-year. A total of 228 foreign trainees were enrolled, benefiting 44 countries. As many as 154 were from Asia, 30 from Africa, 24 from Europe, and 20 from America; there were 220 from developing countries. The males and females were about 60% and 40% respectively. The training courses covered the theory of karst dynamics, karst carbon cycle and addressing climate change, karst rocky desertification control and ecological restoration, karst water resources and water security, karst geological disaster early warning and prevention, karst landscape investigation and development, and other systematic karst knowledge. The training courses improved the social understanding on karst science and promoted in-depth exchanges among participants effectively.

中心面向全球岩溶科技人员（包括学生）、管理人员共计组织各类培训8次，累计培训外籍学员228人次。惠及国家44个，其中亚洲国家154人、非洲国家30人、欧洲国家 24 人、美洲国家 20 人；来自发展中国家的学员 220 人；男女学员占比分别约为 60% 和 40%。培训涵盖岩溶动力学理论、岩溶碳循环与应对气候变化、岩溶石漠化治理与生态修复、岩溶水资源与水安全、岩溶地质灾害预警与防治及岩溶地貌调查与开发等岩溶系统知识，有效提高了社会对岩溶科学的认识，促进了各国学员之间的深入交流。

6. Demonstration sites network
示范基地网

IRCK mainly took the typical karst areas in southwest China for demonstration sites. For example, it set up typical demonstration sites for rocky desertification control and ecological restoration to improve the utilization of epikarst water and groundwater in Yunnan graben basin (such as Luxi, Mengzi, and Jianshui) and Guangxi fengcong depression (such as Huixian in Guilin and Guohua in Pingguo); moreover, it built up a demonstration site for artificially accelerating karstification and carbon cycle by improving soil in Pingguo; it constructed a demonstration site for karst carbon sink research focusing on the water balance and carbon flux within the catchment in Libo of Guizhou; it also set up a demonstration site for cave stalagmite paleoclimate records in Maomaotou Cave of Guilin; it built up a demonstration site for monitoring and early warning of karst collapse and its potential risk early identification in Guangzhou of Guangdong; in addition, it constructed some demonstration sites for sustainable utilization of karst landscape resources in the Lijiang River Basin of Guilin and Wulong of Chongqing. Since southwest China is the best representative for subtropical karst in the world, these demonstration sites in southwest China are important reference for the sustainable development of karst areas within the same or similar climatic zones.

中心主要以中国西南典型岩溶区为示范基地网建设区，在云南断陷盆地（如泸西、蒙自、建水）、广西峰丛洼地（如桂林会仙、平果果化）建设有提高岩溶表层水、地下水资源利用率的石漠化治理与生态修复典型示范区；在广西平果还建设有以改良土壤等方式来人为加速岩溶作用与碳循环的示范区；在贵州荔波建设有以流域水均衡和碳通量研究为主的岩溶碳汇研究示范区；在桂林茅茅头大岩建设有洞穴石笋对古气候记录的示范点；在广东省广州市建设有岩溶地面塌陷地质灾害监测预警和隐患早期识别的示范点；同时，还在桂林漓江流域以及重庆武隆建设有岩溶景观资源可持续利用的示范区。中国西南地区拥有全球最典型的亚热带岩溶地质，在西南地区开展的系列示范对相同和类似气候带的岩溶区可持续发展具有重要参考价值。

7. Monitoring system, modeling system and related mapping system
   监测系统、建模系统和相关制图系统

The monitoring system. With preliminary monitoring system on karst critical zone, established by cooperation with Slovenia, Slovakia, the United States, Brazil, Iran, and Thailand, and supported by such international cooperation projects as IGCP 661 and international geological survey projects, the temperate monsoon climate, subtropical monsoon climate, tropical monsoon climate, and tropical desert climate were well represented. Among them, Thailand and Slovenia karst monitoring stations are fully funded by IRCK, and others by related data sharing. **The modeling system.** IRCK carried out application studies on numerical simulation of karst aquifers based on CFP, MODFLOW, and SWAT models, mainly for karst water cycle and water resources evaluation in China, with ideal results achieved in the Pearl River Basin, Hunan province, Guangxi province, and other typical karst areas. In view of the effective application in heterogeneous karst aquifer, it is planned to promote the research results to ASEAN and Africa in the next operational period. **The mapping system.** IRCK generated 28 maps such as *Karst Map of the World (1:10 million)* and *Serial Maps of Karst Environmental Geology in Southern China and Southeast Asia*, and *Distribution of Areas Prone to Geohazards in Key Areas of five countries on Indo-China Peninsula*, by using ArcGIS and MapGIS to provide effective services for water resources security and important constructions sites selection.

监测系统：依托 IGCP 661 项目、境外地质调查项目等一批国际合作项目，中心与斯洛文尼亚、斯洛伐克、美国、巴西、伊朗和泰国等合作建设了岩溶关键带监测系统，分别代表了温带季风气候、亚热带季风气候、热带季风气候及热带沙漠气候等岩溶环境，其中泰国、斯洛文尼亚岩溶监测站由中心全额资助，其他站点通过数据共享等机制共同开展岩溶关键带监测。**建模系统**：中心开展了基于 CFP、MODFLOW、SWAT 模型的岩溶含水层数值模拟的系列应用研究，主要应用于中国岩溶水循环及水资源评价，在珠江流域以及湖南、广西等典型岩溶区均取得了较为理想的成效，鉴于相关成果在不均一的岩溶含水介质中的有效应用，计划在下一个运行期内推广到东盟、非洲等相关国家和地区。**制图系统**：中心利用 ArcGIS 和 MapGIS 编制了《全球岩溶分布图（1:1000 万）》《中国南部及东南亚地区岩溶环境地质系列图》《中南半岛 5 国重点区地质灾害易发区分布图》等系列图件 28 幅，为各国水资源安全保障、重大工程选址等提供有效服务。

8. Revise and update the website and science popularization and dissemination
修改更新中心网页、科普活动

IRCK replaced the old web pages and set up a new website with a new URL, enriched the content of the new web pages, smoothed browsing, and diversified the information. IRCK organized more than 200 popular science activities with more than 6,500 participants, and publicized karst knowledge through visits, lectures, virtual interaction, teenagers' painting competitions, etc. IRCK organized and edited 51 popular science papers (books), made 3 popular science videos, and developed 2 sets of readings to raise public awareness through the novel and diversified products.

中心更换了旧网页设计，申报了新网址，充实了新网页内容，浏览更加顺畅，信息更加多元化。中心组织了科普活动 200 余次，参与人数 6500 余人，通过参观、讲座、在线互动、青少年绘画比赛等方式宣传岩溶科学知识；中心组织编写了科普读物 51 篇（册），编辑科普视频 3 个，开发研学材料 2 套，以新颖且多样化的科普产品提升公众保护意识。

9. Develop guidelines and criteria on the investigation and research on karst dynamic systems
制定岩溶系统调查和研究的指南和准则

IRCK submitted two international standards proposals to the ISO/TC 319, including *Specification of monitoring technology for karst critical zones*, and *Technical specification of wearable protective clothing for scientific research and exploration enthusiasts into karst cave*, both waiting for voting by members of ISO/TC 319. The natural resources industrial standard *Guidance for karst carbon cycle survey and carbon sink evaluation* formulated by IRCK was approved by the Ministry of Natural Resources and officially implemented in November 2021. IRCK also participated in the preparation of the forestry industrial standard *Technical regulations for monitoring and assessing of karst rocky desertification control project*, and group standard *Code for geological investigation of karst collapse prevention* that was officially implemented on September 1, 2020. In 2022, IRCK completed four draft national standards like *Essential terminology of karst speleology*. IRCK prepared standards actively, providing guidelines and criteria for promoting the comparative study of karst critical zones, karst cave survey, addressing global climate change, rocky desertification control and restoration, biodiversity protection, karst geological disaster prevention and sustainable development, etc.

中心向国际标准化组织岩溶技术委员会提交了《岩溶关键带监测技术》《岩溶洞穴科研及探险用可穿戴防护服技术规范》两项国际标准提案，均已进入ISO投票环节；中心编制的行业标准《岩溶碳循环与碳汇效应评价指南》获得自然资源部批准，于2021年11月对外正式发布；中心还参与编制了行业标准《石漠化治理监测与评价规范》、团体标准《岩溶地面塌陷防治工程勘查规范（试行）》（于2020年9月1日正式实施）；2022年，中心编写完成《岩溶洞穴学基本术语》等4项国家标准征求意见稿。中心积极编制多项标准，为推进岩溶关键带对比研究、岩溶洞穴调查、应对全球气候变化、石漠化治理修复及生物多样性保护、岩溶地质灾害防治和可持续发展等提供了指南和准则。

左图 /Left
Code for geological investigation of karst collapse prevention
《岩溶地面塌陷防治工程勘查规范（试行）》

右图 /Right
Technical regulations for monitoring and assessing of karst rocky desertification control project
《石漠化治理监测与评价规范》

6.2 Contributions to UNESCO
为教科文组织所做的贡献

IRCK carried out related work to support the Main Lines of Action 1, 4, 5 and 6 of natural science priorities (see Chapter 1 for details) in accordance with the UNESCO's prevailing Approved Plan and Budget 38C/5 at the time in which it was approved to be renewed. Moreover, the training and cooperative project made some contributions to the global priorities, "Africa" and "Gender Equality".

中心依据延续批复之时教科文组织批准的计划与预算 38C/5，针对自然科学优先领域工作重点 1、4、5、6（详见第一章）开展了相应工作；通过培训和项目合作，中心还在总体优先事项"非洲优先""性别平等"等领域做出了相应贡献。

6.2.1. Contributions to sectoral programme priorities
中心为部门计划优先事项所做贡献

1. Main Lines of Action 1: Strengthening STI policies, governance and the science-policy-society interface
工作重点 1：加强科技与创新政策、治理及科学 – 政策 – 社会间的互动

Karst carbon sink with carbon peaking and carbon neutrality goals: IRCK is committed to the research of karst carbon cycle based on karst dynamics, with pioneer research results on karst carbon sink effects achieved. It is concluded that the weathering and dissolution of carbonate rocks and the photosynthesis of aquatic plants in karst areas can produce significant short-time scale karst carbon sink effects. In 2020, the Chinese government put forward the carbon peaking and carbon neutrality goals. IRCK's research on karst carbon sink is also linked in related formal documents. The perfect closed loop among karst carbon sink research, the release of national policy and the needs of local governments fully reflects the perfect interface among scientific research, national policies, and social needs.

岩溶碳汇与"双碳"目标：中心致力于开展以岩溶动力学理论为基础的岩溶碳循环研究，在岩溶碳汇效应研究领域取得了引领性的研究成果，总结出碳酸盐岩风化溶解及岩溶区水生植物光合作用可产生显著的短时间尺度岩溶碳汇效应。2020年，中国政府提出"双碳"目标，中心岩溶碳汇研究被正式列入国家有关文件。岩溶碳汇研究与国家文件的发布及地方政府需求之间形成的完美闭环，充分体现了科学研究、国家政策与社会需求之间的完美互动。

Karst rocky desertification control and ecological civilization construction: Different from the common desertification, karst rocky desertification refers to the process in which the bedrock is exposed and the land becomes rocky desert due to the vegetation damage and the soil erosion because of unreasonable human activities in subtropical fragile karst ecological environment. Karst rocky desertification is an ecological deterioration phenomenon common in karst areas of southwest China. IRCK devoted to study the mechanism of karst rocky desertification and the demonstration of its ecological restoration process. The reports of the 18th and 19th National Congresses of the Communist Party of China both attached the significance of the rocky desertification control to ecological protection of natural systems. The rocky desertification control is also important to the ecological civilization construction of the Five-in-One new pattern advocated by the government of China. In 2016, IRCK started to organize and implement the National Key Research and Development Project of China "Evolution of rocky desertification in karst graben basin and its comprehensive control and demonstration", which successfully realized the ecological restoration of karst rocky desertification in Yunnan graben basins. A perfect science-policy-society interface is thus indicated from the appearance of the concept of rocky desertification, to the concept of rocky desertification control mentioned several times in the national reports, and then the wide promotion and application of related technology in southwest China.

岩溶石漠化治理与生态文明建设：岩溶石漠化有别于常见的荒漠化，是指在亚热带脆弱的岩溶生态环境下，由于不合理的人类活动作用导致植被遭受破坏、土壤严重流失而引起的基岩裸露，土地呈现荒漠化景观的过程。岩溶石漠化是中国西南岩溶区常见的生态恶化现象。中心致力于岩溶石漠化发生机理及其生态修复过程的研究与示范，在中国政府第十八次、第十九次全国代表大会上的报告中，均特别强调了石漠化治理对自然系统生态保护的重要性，开展石漠化治理也是中国政府倡导的"五位一体"中生态文明建设的重要支撑。2016年，中心组织实施了中国国家重点研发计划项目"喀斯特断陷盆地石漠化演变及综合治理技术与示范"，在云南断陷盆地成功实现了岩溶石漠化的生态修复。从石漠化概念的出现，到石漠化治理理念进入国家报告，再到石漠化治理技术在中国西南地区的广泛推广与应用，展示了科学研究、国家政策与社会需求之间的完美互动。

Targeted poverty alleviation, rural revitalization and karst scientific research: From targeted poverty alleviation proposed in 2013 to the rural revitalization proposed in 2017 by the government of China, IRCK has carried out systematic scientific research to support national policies, including the research on the development and utilization of water resources in karst areas, the ecological restoration of karst rocky desertification areas, and the development and utilization of karst geological landscapes. IRCK helped to solve the principle demands of water use for poverty-stricken areas through water development, helped to solve the problems of water leakage and soil loss faced by agricultural development in poverty-stricken areas through ecological restoration, and helped to solve the problem of sustainable economic benefits in poverty-stricken areas through the development of geological eco-tourism. IRCK made serial great efforts to respond to national policies and social needs positively.

精准扶贫、乡村振兴与岩溶科学研究：从中央领导人于2013年提出精准扶贫到2017年提出乡村振兴，中心开展了系统性的科学研究用以支撑国家政策，包括岩溶区水资源的开发利用研究、岩溶石漠化区的生态修复及岩溶地质景观的开发利用等，通过水资源开发解决贫困区基本用水问题，通过生态修复解决贫困区农业开发面临的水土漏失难题，通过开发地质生态旅游解决贫困区可持续经济效益难题。中心的系列科学研究是对国家政策和对社会需求的积极响应。

2. Main Lines of Action 4: Fostering international science collaboration for earth systems, biodiversity, and disaster risk reduction

工作重点4：促进地球系统、生物多样性和降低灾害风险方面的国际科学合作

The International Big Scientific Plan on Global Karst launched by IRCK aims to promote international scientific cooperation in different fields of karst, including the earth system (research on karst critical zone), biodiversity (research on karst ecosystems), and disaster risk reduction (research on early warning and prevention technologies of karst geological disasters). According to the expected results in 38C/5, IRCK has made the following contributions in promoting international scientific cooperation in the earth system and promoting the number of member states of global geoparks.

中心启动的"全球岩溶"国际大科学计划旨在推动岩溶全领域的国际科学合作，包括地球系统（岩溶关键带研究）、生物多样性（岩溶生态系统研究）和降低灾害风险（岩溶地质灾害预警与防治技术研究）。按照38C/5中各项预期成果，中心在促进地球系统国际科学合作和推进全球地质公园的会员国数量方面做出了以下贡献。

Implementation of IGCP 661: The IGCP 661 project "Processes, Cycle and Sustainability of the Critical Zone in Karst Systems" (2017-2021) has clarified the structural characteristics of karst critical zone and revealed the characteristics of ecological hydrological process and carbon cycle of karst critical zones. It is a powerful supplement to promote the study of earth critical zones, and also an effective promotion of the international scientific cooperation of the earth system. The project set 10 leader or co-leaders, with 6 co-leaders from developing countries, accounting for 60%.

实施 IGCP 661 国际地球科学计划：IGCP 661 项目"岩溶关键带物质能量循环过程及可持续性研究"（2017~2021 年）厘清了岩溶关键带结构特征，揭示了岩溶关键带生态水文过程及碳循环特征，是推进地球关键带研究的有力补充，有效促进了地球系统的国际科学合作。项目设有 10 位主席或联合主席，其中有 6 位来自于发展中国家，占比达 60%。

Assisting Thailand in applying for its first UNESCO Global Geopark: In 2018, at the invitation of Thailand, IRCK assisted Thailand in successful application for the first UNESCO Global Geopark in Thailand – Satun UNESCO Global Geopark. IRCK went to Satun of Thailand to provide technical guidance before its evaluation. The successful application of Satun Geopark means that Thailand has become a member of the UNESCO Global Geopark network. IRCK has made some contributions to increase the number of Member States with Global Geoparks set as a performance indicator by UNESCO.

协助泰国申报第一个世界地质公园：2018 年，应泰方邀请，中心协助泰方成功申报了泰国境内首个世界地质公园——沙敦世界地质公园。中心赴泰国沙敦给予了现场迎检前的技术指导。沙敦世界地质公园的成功申报意味着泰国成为世界地质公园网络会员国，中心的工作成功助力教科文组织实现扩大世界地质公园网络的预期目标。

3. Main Lines of Action 5: Strengthening the role of ecological sciences and biosphere reserves

工作重点 5：强化生态科学和生物圈保护区的作用

Implement National Key Research and Development Project of China "Evolution of rocky desertification in karst graben basins and its comprehensive control and demonstration": Since 2016, IRCK has implemented National Key Research and Development Project of China "Evolution of rocky desertification in karst graben basin and its comprehensive control and demonstration", focusing on the evolution mechanism, ecological restoration, and ecological industry construction of rocky desertification in Yunnan graben basin in China. Two sets of vegetation restoration technologies that can be effectively applied to karst ecosystem in southwest China have been successfully explored, i.e. rapid restoration of vegetation by interplanting indigenous trees, shrubs, and grass, as well as transplanting soil seed bank to explore the restoration of natural vegetation community. About 1,992 mu rocky desertification area was rehabilitated, with the vegetation coverage increased by 31.6%. IRCK started with the evolution law of karst ecosystem and reached the rapid restoration through appropriate human intervention, which serves as an effective practice of karst ecological science with important significance for extensive promotion.

实施国家重点研发计划"喀斯特断陷盆地石漠化演变及综合治理技术与示范"：中心自2016年开始实施国家重点研发计划"喀斯特断陷盆地石漠化演变及综合治理技术与示范"，着重开展中国云南断陷盆地石漠化的演变机制、生态修复和生态产业构建，成功探索了两套可有效应用于西南岩溶生态系统的植被修复技术，即土著乔-灌-草套种抚育植被快速修复和土壤种子库移植探索仿自然的植被群落修复。修复石漠化面积1992亩，植被覆盖度提升31.6%。中心从岩溶生态系统的演化规律着手，通过适度人为干预，实现快速修复。这是岩溶生态科学的一次有效实践，具有广泛的推广意义。

Cooperated with the Chinese National Committee for Man and the Biosphere Programme to edit *Karst Geology and Ecosystem — Karst Science Popularization in China*: IRCK cooperated with the Chinese National Committee for Man and the Biosphere Programme (MAB) to edit a science popularization book focusing on karst geology and karst ecosystem, publicize the characteristics of karst geology and karst ecosystem, and how to protect karst ecosystem effectively by considering its geological characteristics. The book, using popular science words, has promoted the development of karst ecological science effectively.

与中国人与生物圈国家委员会合作编制《岩溶与生态——岩溶科普在中国》：中心与中国人与生物圈（MAB）国家委员会合作编制了聚焦岩溶地质与岩溶生态系统的科普图书，宣传岩溶地质的特征、岩溶生态系统的特色及如何利用岩溶地质特征有效保护岩溶生态系统。该书从科普视角有力地推进了岩溶生态科学的发展。

4. Main Lines of Action 6: Strengthening freshwater security
工作重点 6：加强淡水安全

Developed four modes for the utilization of karst water resources: According to the karst hydrogeological and the developing characteristics of underground rivers of different areas, IRCK developed four modes for sustainable utilization of groundwater resources: blocking the conduits of underground rivers to form reservoirs in karst hilly depressions, regulating and storing epikarst water in karst peak-cluster areas, pumping in hilly valleys, and damming in graben basins, which supported the water supply in water-short karst mountainous areas powerfully.

研发四种岩溶水资源开发利用模式： 根据岩溶水文地质、地下河发育特征，因地制宜地研发了溶丘洼地区地下河堵洞成库模式、峰丛山区表层岩溶水调蓄模式、丘陵谷地储水构造抽水调节模式和断陷盆地雍水调度模式四种地下水可持续开发利用模式，有力支撑了岩溶石山缺水区的水资源供应。

Carried out research on karst water security of China: IRCK carried out the ecological restoration of large karst springs in northern China, the prevention and control of karst water polluted by acid "old kiln water" in closed coal mines, and risk assessment of pollution degree of inorganic and organic indicators of karst underground river systems in southwest China, providing scientific support for safe development and utilization of freshwater resources in karst areas in China.

开展了中国岩溶水安全研究： 中心开展中国北方岩溶大泉泉水复流生态修复、闭坑煤矿酸性"老窑水"对岩溶水的污染防治、中国西南地区岩溶地下河系统的无机指标和有机指标污染程度风险评价，为中国岩溶区淡水资源的安全开发与利用提供了科学支撑。

6.2.2 Contributions to global priorities
中心为总体优先事项所做贡献

Africa: IRCK invited African scientists and technicians to join the International Big Scientific Plan on Global Karst actively, signed a cooperation agreement with the National University of Science and Technology of Zimbabwe, establishing a karst hydrogeological research team with Mr. Innocent Muchingami as the leading scientist. IRCK cooperated with the Zimbabwean team to apply for the GEO Pilot Initiative "Earth Observations for Global Typical Karst". IRCK carried out bilateral cooperation on karst water resources management with South Africa according to the agreement signed previously, and establishing a karst water resources research team with Mr. Harrison Pienaar from the Council for Scientific and Industrial Research (CSIR) as the leading scientist. Through international training courses, IRCK has trained 30 African students in total. In 2021, IRCK built up a bridge with the young scientists working for karst in the Republic of Congo through the training course, planning to apply for the International Geoscience Program jointly. In addition, IRCK hosted three China-Africa Water Resources Dialogues to discuss the sustainable management of water resources in developing countries. IRCK has made great contributions to the development of karst science in Africa through cooperative projects, training, and seminars.

非洲优先： 中心积极邀请非洲科技人员加入"全球岩溶"国际大科学计划，与津巴布韦国立科技大学签署合作协议，形成了以因纳森·穆钦加米先生为领军人才的岩溶水文地质研究团队，中心与该团队合作申请 GEO 试点项目"全球典型岩溶区观测"。中心与南非围绕前期签署的合作协议开展岩溶水资源管理双边合作，形成了以南非科学和工业研究委员会专家哈里森·皮纳尔先生为领军人才的岩溶水资源研究团队。中心通过国际培训，累计培训非洲学员 30 名。2021 年，中心通过培训，与刚果共和国岩溶科研人员搭建起合作桥梁，拟共同申请国际地球科学计划项目。此外，中心共计主办中非水资源论坛 3 次，探讨发展中国家水资源可持续管理。中心通过项目合作、人员培训、举办会议等方式为非洲岩溶科学发展做出了有力贡献。

Chapter 6　Summary

Gender Equality: IRCK has enrolled 91 female scientists and technicians as trainees in the second six years, accounting for about 40% of the total trainees. The proportion of female trainees has increased obviously. The IGCP 661 has attracted 14 females (30% of the total) to participate in relevant work, including one from Slovakia as a co-leader. Through training and project cooperation, IRCK has made some contributions to improving the research capacity of females from developing countries.

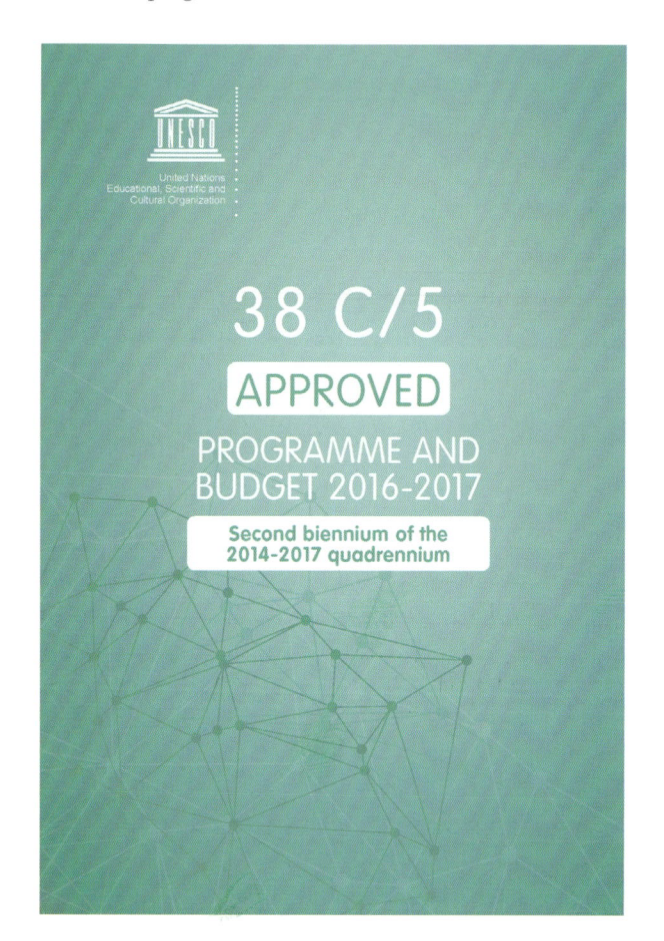

性别平等：中心在第二个六年历程内累计培训女性科技人员 91 名，约占培训人员总量的 40%，中心培训女性科技人员比例显著上升；IGCP 661 项目累计吸纳 14 名女性科技人员参与相关工作，其中 1 名来自斯洛伐克的女性科技人员担任项目联合主席，女性参与人员约占科技人员总量的 30%。中心通过培训和项目合作，为提升发展中国家女性科技人员的科研水平做出了贡献。

Approved Programme and Budget 38C/5
教科文组织批准的计划与预算 38C/5

6.3 Contributions to UN 2030 Agenda for Sustainable Development
为联合国 2030 年可持续发展议程所做的贡献

IRCK has made contributions to the United Nations 2030 Agenda for Sustainable Development, especially for the goals of Clean Water and Sanitation (Goal 6), Industry, Innovation and Infrastructure (Goal 9), Sustainable Cities and Communities (Goal 11), Climate Action (Goal 13), Life on Land (Goal 15), and so on.

中心为联合国 2030 年可持续发展议程做出了一系列贡献，尤其是在清洁饮水和卫生设施（目标 6）、产业、创新和基础设施（目标 9）、可持续城市和社区（目标 11）、气候行动（目标 13）、陆地生物（目标 15）等目标领域，取得了突出成果。

Goal 6: Clean Water and Sanitation
目标 6：清洁饮水和卫生设施

The four water resource utilization modes proposed by IRCK (see Section 6.2.1 "Main Lines of Action 6: Strengthening freshwater security") were widely used in southwest China. More than 200 demonstration projects for the development and utilization of karst water resources were implemented, with more than 8,000 wells drilled successfully, directly alleviating the drinking water difficulties of 6 million people in poverty-stricken karst areas with serious water shortage problems, and guaranteeing clean drinking water sources for more than 20 million people. IRCK also provided suggestions to control the karst aquifers polluted by "old kiln water" in mining areas in northern China, and carried out investigation and research on the pollution of karst underground rivers in southwest China. The work is a significant support for guaranteeing clean drinking water in karst areas.

中心提出的四种水资源利用模式（详见 6.2.1 节"工作重点 6：加强淡水安全"）在中国西南地区被广泛应用，共实施岩溶水资源开发利用示范工程 200 多处，勘探成井 8000 多眼，直接解决岩溶石山严重缺水地区和贫困地区 600 万人饮用水困难，为 2000 多万人提供了清洁的饮用水源保障。中心还为中国北方矿区"老窑水"污染岩溶含水层提供治理建议，开展了西南岩溶地下河污染状况调查研究，为保障岩溶区清洁饮水提供了科技支撑。

Goal 9: Industry, Innovation and Infrastructure
目标 9：产业、创新和基础设施

Geological disaster early warning and prevention: IRCK established a perfect dynamic monitoring and early warning system for karst collapse, zoning the prone areas for karst collapse in China, and made demonstration in southern China. IRCK provides effective scientific and technological guarantee for people living in karst areas to resist related risks and disasters.

地质灾害预警和预防：中心建立了完善的岩溶塌陷动态监测和预警系统，划分了中国岩溶易发区，在中国南方开展系列示范。中心为岩溶区人民抵御相关风险和灾害提供了行之有效的科技保障。

Goal 11: Sustainable Cities and Communities
目标 11：可持续城市和社区

Conservation of World Natural Heritage sites: IRCK supported the successful application of Wulong World Natural Heritage Site, Jinfo Mountain World Natural Heritage Site, and Guilin World Natural Heritage Site. During the 2nd operation phase, IRCK has helped these sites to carry out science popularization and strengthen their protection and management. In 2018, Wulong World Natural Heritage Site was selected as the Practice and Innovation Base of "Lucid waters and lush mountains are invaluable assets". In 2020, it was selected as the National Ecological Civilization Construction Demonstration Area, acting as a typical example of perfect integration of natural heritage protection, ecological civilization construction, and tourism development.

自然遗产地保护：中心曾支撑武隆世界自然遗产地、金佛山世界自然遗产地、桂林世界自然遗产地的成功申报；并在二期运营期内，助力各世界遗产地开展科学普及、加强遗产地保护与管理。武隆世界遗产地于 2018 年被评选为绿水青山就是金山银山的实践创新基地，2020 年被评选为国家生态文明建设示范区，成为将自然遗产地保护、生态文明建设和旅游发展完美融合的典型范例。

Goal 13: Climate Action
目标 13：气候行动

IRCK carried out research on karst carbon sink and put forward four ways to increase carbon sink artificially, providing new artificial approaches for carbon emission reduction in karst areas. IRCK also carried out the research on travertine which could be used as another paleoclimate record, also on the stalagmites for paleoclimate change. IRCK endeavored to predict the climate change trend in future through the research on paleoclimate and its evolutionary trend.

中心开展了岩溶碳汇效应研究，提出了人工增汇的四大途径，为岩溶区碳减排增添新的人工干预手段；还开展以钙华沉积物为媒介的气候变化记录研究，以及以石笋为介质的古气候变化研究，通过古气候及其演化趋势研究，预测未来气候的变化趋势。

Goal 15: Life on Land
目标 15：陆地生物

IRCK made outstanding contributions to the protection of terrestrial organisms by researching on ecological restoration of rocky desertification and karst wetland ecology. IRCK is committed to repairing degraded karst ecosystems, protecting fragile karst ecological environment, and protecting the biodiversity of special karst habitats. Through training and scientific popularization, IRCK enhanced the capacity of practitioners, improved relevant protection measures, and enhanced public awareness of protection, realizing protection from theory to practice, and improving the relevant literacy of the general public and practitioners.

中心通过开展石漠化生态修复研究、岩溶湿地生态研究等为保护陆地生物做出了突出贡献。中心致力于修复退化的岩溶生态系统，保护脆弱的岩溶生态环境和特殊岩溶生境的生物多样性；通过培训和科学普及工作，提升专业人员的保护力度，完善相关保护措施，提升公众保护意识。中心实现了从理论到实践的保护之路，提升了普通大众及专业人员的相关素养。

6.4 Contributions to the Member States
为会员国所做的贡献

Over the past six years, IRCK has devoted to building an active, efficient, and influential international platform, aiming to provide effective services to member states and promote the development of karst science jointly. In general, in the second six years, IRCK has made serial contributions to member states through international training courses and academic exchanges, project cooperation and social services.

中心六年来致力于打造活跃、高效、富有影响力的国际平台，旨在为各会员国提供有效服务，共同推进岩溶科学发展。总体来说，在第二个六年历程中，中心通过国际培训与学术交流、项目合作与社会服务等为会员国做出了系列贡献。

International training and academic exchange: Over the past six years, IRCK has hosted 18 important international and domestic conferences, providing an academic stage for karst science exchange for nearly 50 countries. Over the past six years, IRCK has trained 228 karst scientists (226 from the member states), technicians, and managers from 44 countries (43 are member states), including 15 Asian countries, 13 African countries, 12 European countries, 4 American countries(3 are member states), 37 developing countries and one small island country, namely Jamaica. The training improved the theoretical level on karst science of member states effectively and promoted the long-term development of karst science worldwide. The training helped excellent trainees to be promoted. For example, Mr. Eko Haryono, the trainee from Indonesia, was elected as the chairman of the Karst Commission under the International Geographic Union; and Alena Gessert, the trainee from Slovakia, was elected as the secretary of the European Speleological Federation.

国际培训与学术交流：中心六年来共主办了18次重要国际国内会议，为近50个国家提供了岩溶科学交流的学术舞台；中心六年来共计培训了44个国家（43个为会员国）的岩溶科技人员和管理人员228人次（226人次为会员），其中亚洲国家15个、非洲国家 13 个、欧洲国家 12 个、美洲国家 4 个（3 个为会员国），发展中国家 37 个，小岛屿国家 1 个，即牙买加。通过培训，有效提升了会员国岩溶科学理论水平，在全球范围内推进了岩溶科学长远发展。印度尼西亚学员艾可·哈约诺培训后当选为国际地理联合会岩溶专业委员会主席，斯洛伐克学员阿琳娜·吉斯特培训后当选为欧洲洞穴联合会秘书长。

Project cooperation: IRCK has established good cooperation relations with Southeast Asian countries (Cambodia, Indonesia, Laos, Malaysia, Myanmar, the Philippines, Thailand, and Vietnam), and jointly implemented five international cooperation projects under the framework of Global Karst. IRCK has jointly implemented a number of multilateral or bilateral cooperation projects with European, American, and central Asian countries, including three China-Slovenia bilateral cooperation projects, the China-Slovakia bilateral cooperation project, and IGCP 661, a multilateral cooperation project. Through multilateral cooperation with 26 countries (see Section 3.1 for details), IRCK has established a karst critical zone monitoring network with Thailand, Slovenia, Slovakia, the United States, Brazil, Spain, and Iran, promoted the comparative study of karst geology and the research on earth system in karst critical zone jointly. IRCK has established a good cooperative relationship with Africa, mainly conducting bilateral and multilateral cooperation with Zimbabwe and South Africa for the International Big Scientific Plan on Global Karst, and plans to jointly apply for the International Geoscience Program with the Republic of Congo.

项目合作： 中心与东南亚诸国（柬埔寨、印度尼西亚、老挝、马来西亚、缅甸、菲律宾、泰国、越南）建立了良好的合作关系，在"全球岩溶"国际大科学计划的框架下共同实施了 5 项国际合作项目；中心与欧洲、美洲及中亚国家共同实施了多个多双边合作项目，包括 3 项中国－斯洛文尼亚双边合作项目，1 项中国－斯洛伐克双边合作项目，1 项 IGCP 多边合作项目，中心通过与 26 个国家（详见 3.1 节）开展多边合作，与泰国、斯洛文尼亚、斯洛伐克、美国、巴西、西班牙、伊朗等国构建岩溶关键带监测网，共同推进岩溶地质对比研究及岩溶关键带地球系统科学研究；中心与非洲建立了良好的合作关系，主要与津巴布韦、南非开展"全球岩溶"国际大科学计划下的双多边合作，并计划与刚果共和国联合申报国际地球科学计划项目。

Social services: The Karst Technical Committee of the International Organization for Standardization operated by IRCK has 21 observing members and 8 participating members currently, all of which are UNESCO member states. The participation of these participating members and observing members has greatly promoted the establishment and application of karst standards required by relevant member states. In addition, IRCK assisted Thailand in establishing the first UNESCO Global Geopark, helped Thailand to make a breakthrough of the UNESCO Global Geopark, and promoted the sustainable tourism industry development in the coastal karst area of southern Thailand effectively.

社会服务：中心申请的国际标准化组织岩溶技术委员会目前共有21个观察国和8个参与国，均为教科文组织会员国，这些参与国和观察国的加入对推动相关会员国所需岩溶标准的建立与应用起到了极大促进作用。此外，中心协助泰国建立其第一个世界地质公园推动了泰国突破世界地质公园壁垒，有力推进了泰南滨海岩溶区可持续旅游产业发展。

With the strong support of UNESCO, IRCK has become a hub linking the member states, also an important stage for the member states to seek karst information and karst cooperation.

在教科文组织的大力支持下，中心成为连接各会员国的枢纽，成为各会员国寻求岩溶资讯、岩溶合作的重要舞台。

Postscript

Currently, the new base of IRCK has been fully completed. Meanwhile, IRCK has started the construction of experimental sites in the new base: Karst Carbon Cycle and Carbon Sources & Sinks Experimental Site, Karst Ecosystem Experimental Site, Karst Groundwater Cycle and Evolution Simulation Experimental Site, and Large Scale Physical Model Experimental Site of Karst Collapse. After officially being put into use, the new base will play an important role in international exchange, integrating academic discussion, guest research, indoor experiments, and field experiments, which will facilitate the communication, cooperation, and exchange of global karst scientists and technicians. The third phase operation of IRCK will be committed to creating a more active and influential international cooperation platform, making important contributions to promoting the development of karst geological sciences and the sustainable socio-economic development of karst areas.

后记

目前，中心新基地已正式竣工，并正式启动了试验场建设——岩溶作用碳循环与源汇关系试验场、岩溶生态试验场、岩溶地下水循环与演化模拟试验场、岩溶塌陷大型物理模拟试验场。中心新基地正式投入使用后将被打造成国际交流的重要场所，将学术研讨、客座研究、室内实验、野外试验集于一体，方便全球岩溶科技人员开展沟通、协作与交流。中心第三期运营将致力于打造更加活跃、更富影响力的国际合作平台，为推动岩溶地质科学发展和岩溶区社会经济可持续发展做出重要贡献。

后记

New base of IRCK

中心新基地

国际岩溶研究中心第二个六年历程